Fumigants: Environmental Fate, Exposure, and Analysis

A C S   S Y M P O S I U M   S E R I E S   **652**

Fumigants

Environmental Fate, Exposure, and Analysis

EDITORS

James N. Seiber
University of Nevada—Reno

James A. Knuteson
DowElanco

James E. Woodrow
University of Nevada—Reno

N. Lee Wolfe
U.S. Environmental Protection Agency

Marylynn V. Yates
University of California—Riverside

S. R. Yates
Agricultural Research Service, U.S. Department of Agriculture

Developed from a symposium sponsored
by the Division of Agrochemicals

American Chemical Society, Washington, DC

Library of Congress Cataloging-in-Publication Data

Fumigants: environmental fate, exposure, and analysis / editors, James N. Seiber . . . [et al.].

    p.   cm.—(ACS symposium series, ISSN 0097–6156; 652)

"Developed from a symposium sponsored by the Division of Agrochemicals at the 210th National Meeting of the American Chemical Society, Chicago, Ill., August 20–25, 1996."

Includes bibliographical references and indexes.

ISBN 0–8412–3475–2

1. Fumigants—Environmental aspects—Congresses.

    I. Seiber, James N., 1940–   . II. American Chemical Society. Division of Agrochemicals. III. American Chemical Society. Meeting (210th: 1996: Chicago, Ill.) IV. Series

TD196.F86F86   1996
628.5′29—dc21   96–44412
                                                        CIP

This book is printed on acid-free, recycled paper.

Foreword

THE ACS SYMPOSIUM SERIES was first published in 1974 to provide a mechanism for publishing symposia quickly in book form. The purpose of this series is to publish comprehensive books developed from symposia, which are usually "snapshots in time" of the current research being done on a topic, plus some review material on the topic. For this reason, it is necessary that the papers be published as quickly as possible.

Before a symposium-based book is put under contract, the proposed table of contents is reviewed for appropriateness to the topic and for comprehensiveness of the collection. Some papers are excluded at this point, and others are added to round out the scope of the volume. In addition, a draft of each paper is peer-reviewed prior to final acceptance or rejection. This anonymous review process is supervised by the organizer(s) of the symposium, who become the editor(s) of the book. The authors then revise their papers according to the recommendations of both the reviewers and the editors, prepare camera-ready copy, and submit the final papers to the editors, who check that all necessary revisions have been made.

As a rule, only original research papers and original review papers are included in the volumes. Verbatim reproductions of previously published papers are not accepted.

ACS BOOKS DEPARTMENT

Contents

Preface

FUMIGANTS ARE VITAL AGENTS in the production of many food crops, particularly the high-value crops such as strawberries and grapes, which are susceptible to nematodes and other soil-borne pests. They are also used, even required, for fumigating fruits, grains, and spices destined for export.

Fumigants are, however, mobile compounds in the environment and warrant exceptional safeguards in terms of application technology, minimizing worker exposure, and preventing movement to air and groundwater. This is a particular challenge because fumigants are volatile compounds and many are fairly soluble in water. They are also toxic chemicals and thus pose risks to applicators and field workers, and to a lesser extent, to those who live in the vicinity of fumigant operations.

In the past, society has banned agricultural chemicals that are too mobile and toxic to guarantee safety to people and the environment. The fumigant class has been hit hard in this regard. Dibromochloropropane (DBCP) and ethylene dibromide (EDB) have, for example, been banned because of a combination of mobility (groundwater contamination) and toxicity (potential carcinogenicity).

Of the remaining fumigants, virtually all (methyl bromide, 1,3-dichloropropene (Telone), and ethylene oxide) have been threatened with severe limitations, including outright bans. Fortunately, there may be time to learn more about them so that a ban is not necessary. To avoid a ban, we must be able to control exposures as well as air and groundwater contamination.

The focus of the symposium on which this volume is based was to share information on what is known about fumigants and how we can use this information to prevent exposure and adverse effects, while still enjoying the benefits fumigants afford in food production and pest control. The Symposium was presented at the 210th National Meeting of the American Chemical Society and was sponsored by the Division of Agrochemicals, in Chicago, Illinois, from August 20–25, 1996. The chapters in this book cover several fumigants and include information on their environmental fate, properties, emissions, downwind behavior and exposure, and analytical methods.

The Symposium organizers wish to thank the authors and the American Chemical Society for their time and dedication to this publication.

JAMES N. SEIBER
Center for Environmental Sciences and Engineering
and Department of Environmental and Resource Sciences
Mail Stop 199
University of Nevada
Reno, NV 89557–0187

JAMES A. KNUTESON
DowElanco
9330 Zionsville Road
Indianapolis, IN 46268–1054

JAMES E. WOODROW
Center for Environmental Sciences and Engineering
and Department of Environmental and Resource Sciences
Mail Stop 199
University of Nevada
Reno, NV 89557–0187

N. LEE WOLFE
Ecosystems Research Division
National Exposure Research Laboratory
U.S. Environmental Protection Agency
960 College Station Road
Athens, GA 30605–2700

MARYLYNN V. YATES
Department of Soil and Environmental Sciences
University of California
2208 Geology
Riverside, CA 92521

S. R. YATES
Soil Physics and Pesticide Research Unit
U.S. Salinity Laboratory
Agricultural Research Service
U.S. Department of Agriculture
450 West Big Springs Road
Riverside, CA 92507

July 23, 1996

Chapter 1

Health and Environmental Concerns Over the Use of Fumigants in Agriculture: The Case of Methyl Bromide

Puttanna S. Honaganahalli and James N. Seiber

Center for Environmental Sciences and Engineering and Department of Environmental and Resource Sciences, Mail Stop 199, University of Nevada, Reno, NV 89557–0187

With the discovery of oceans and soils as net sinks for methyl bromide (MeBr), a decrease in ozone depletion potential from 0.7 to about 0.45 and overall lifetime from 2.0 to between 0.8 and 1.0 yrs. have been estimated. Further, MeBr has tested negative as a carcinogen. Stringent fumigation rules, at least in the United States, have been effective in reducing exposure. This new information on MeBr weakens the argument for an outright ban of MeBr and suggests that, with better management practices, the chemical could continue to be used beneficially. MeBr is predominantly a naturally occurring compound and anthropogenic sources represent about 25% (+/-10%) of the total emissions.

Methyl bromide (MeBr) has been used as a fumigant since the 1940s with production peaking at 71,500 metric tons in 1992 *(1)*. After the loss of the effective fumigant ethylene dibromide in the early 1980's, MeBr began to be used more widely and became the fumigant of choice because of its effectiveness against a wide spectrum of pests including arthropods, nematodes, fungi, bacteria and weeds (see Chapter 2). Agriculture is the major consumer of synthetic MeBr, followed by structural use and the chemical industries. It is used in strawberry cultivation, and for vegetables such as tomatoes, peppers, and eggplants. Vineyards are fumigated with MeBr before replanting of new vines. It is used to fumigate the soil in fruit and nut orchards. Other important uses of MeBr include the production of tree seedlings for reforestation, and in the strawberry nursery industry to keep the young plants free from soil borne diseases.

Post harvest fumigation now depends mostly on MeBr. It is the primary fumigant recommended for use in the movement of susceptible commodities from a quarantined area containing an introduced pest. Recognizing this fact, the signatories to the Montreal Protocol exempted quarantine uses from regulation. MeBr production has been frozen at

0097–6156/96/0652–0001$15.00/0
© 1996 American Chemical Society

the 1991 levels and it is scheduled to be banned from usage on January 1, 2001, in the USA. As the understanding of the role of chlorine and bromine in the destruction of ozone in the midlatitude lower stratosphere is improving, the anthropogenic sources of chlorine and bromine compounds are under scrutiny as contributors to the global ozone depletion problem. Chlorofluorocarbons (CFCs), for example, are now being replaced with hydrochlorofluorocarbons (HCFCs) and hydrofluorocarbons (HFCs) which are more reactive in the atmosphere and thus do not survive to the stratosphere for release of Cl atoms. Halons, a class of brominated organic compounds formerly used as fire retardant, are no longer produced. There are no other large uses of brominated organic compounds other than MeBr, so that MeBr is being specially scrutinized for its role in ozone loss in the midlatitude lower stratosphere. Because it has a fairly long atmospheric lifetime it is considered to be capable of diffusing into the stratosphere where it could undergo photodissociation to release bromine atoms which can then react with ozone. Thus MeBr has been identified as an ozone depleting substance *(2)*.

According to the U.S. Clean Air Act any compound having an ozone depleting potential (ODP) ≥ 0.2 is classified as an ozone depleting substance (ODS). The ODP of a compound is dependent on the amount emitted and its ability to release ozone depleting breakdown products following diffusion to the stratosphere and on the lifetime of the compound in the atmosphere. Based on somewhat incomplete scientific information, the ODP of MeBr was first estimated as 0.7 *(2)* placing it clearly in the ODS group. The Montreal Protocol and the subsequent Copenhagen Amendments and Adjustments relating to the phaseout of MeBr production and consumption were based on this incomplete knowledge. The Protocol recommends phaseout of MeBr by 2010 in industrialized countries, and a freeze in consumption by 2002 in developing countries. In the U.S., the Clean Air Act Amendment requires a phaseout of MeBr by 2001 because of the ODP. The estimation of the annual MeBr addition to the atmosphere, the overall lifetime of the molecule, the amount of ozone depletion caused exclusively by MeBr, and the general sense of environmental conservatism prevailing at the time were some of the other important factors that were responsible for this extreme measure. It was estimated that over 50% of MeBr produced for fumigation is emitted into the atmosphere which amounts to an annual addition of 3pptv or 51Gg of MeBr. This is based on broad estimates of emission factors for different areas of the world, with their diverse soil types and application practices. Hence these estimates of losses must be viewed with caution.

The overall lifetime of MeBr was estimated as 2.0 yrs. *(2)* based only on its reaction with OH radicals in the atmosphere. Oceanic, soil, and plant foliage sinks were not known and not addressed. The estimation that at current levels of MeBr emissions bromine would account for 5-10% of the total global ozone loss is, once again, a rough estimate since it is dependent on the rate of formation of HBr in the stratosphere about which there is a major uncertainty *(2)*. Although the science of MeBr in the environment was in a nascent state and unable to provide definitive answers to many questions, a sense of conservatism guided the recommendations for elimination of MeBr.

Since the Montreal Protocol first addressed these issues in 1992 and in its subsequent meetings (Copenhagen, 1994 and Vienna, 1995), significant new information has been uncovered which might support a re-evaluation of the phaseout of MeBr. This is especially important considering the huge economic losses that the phaseout could bring to a country's agricultural sector and economy at large *(3)*. New information regarding the sources, emissions, fate and transport, and sinks have lowered the estimated lifetime of MeBr in the atmosphere to between 0.8 yr. *(4)* and 1.0 yr. *(5)* and the ODP to 0.45. The focus of this chapter is therefore to provide an update on the issue of MeBr and its effect on ozone depletion, and to relate some of the experiences with MeBr to other fumigants.

Sources
MeBr has both natural and anthropogenic sources which complicates the estimation of global burden, lifetime and ODP. Adding to the complication is the presence of such sources as biomass burning which has both natural and anthropogenic origins.

Natural Sources
   Biomass Burning: The WMO 1995 *(6)* report suggests a range of 10-50 Ggy^{-1} and recent work puts the best estimate at 20 Ggy^{-1}*(7)*. There is a large amount of uncertainty in this term because the proportion of MeBr emitted depends on the temperature of the flame and, to some degree, upon the composition and location of the vegetation. Global extrapolation of limited measurements increases the uncertainty of this estimate *(4)* even further. Estimating the definite amounts caused by each source and then quantifying the MeBr released from these separate sources have proven difficult.
   Oceanic Source: Another source with much uncertainty is the ocean. Complication arises from the fact that the ocean is both a source and a sink. The WMO 1995 *(6)* report puts the source at 90 Ggy^{-1} with a possible range of 60-160 Ggy^{-1} based mainly upon the data of Khalil et al. *(8)* and the review of Singh and Kanakidou *(9)*. NOAA *(10)* places the oceanic source at 45 Ggy^{-1}. This subject is receiving much research interest and hence the source and sink parameters and resulting lifetime estimates are undergoing frequent revision. The current best estimate is 60 Ggy^{-1} with a probable range of 30-100 Ggy^{-1} *(5)*.

Anthropogenic Sources
Anthropogenic sources are primarily emissions from pre-plant soil, post-harvest quarantine and structural fumigations, and automobile emissions. Industrial production peaked in 1992 at 71,500 metric tons (71.5 Gg). Of this production 77% was used for pre-plant soil fumigation, 12% for post-harvest commodity fumigation and structural fumigation while 6% was used in chemical intermediates *(11)*.
   Automobile Sources: Automobiles using leaded gasoline, mostly in developing countries, with ethylene dibromide as an additive emit small but measurable amounts of MeBr. The WMO 1995 *(6)* report cites two studies. One places the emissions between 0.5-1.5 Ggy^{-1} while the other places the emissions between 9-22 Ggy^{-1}, an order of magnitude higher than the first. Thus there is a large disagreement in extrapolations of these data and the emissions range from 0.5-22 Ggy^{-1}.
   Post-harvest And Structural Fumigation: Emissions from post-harvest (See Chapter 14) and structural fumigations are well quantified. These emissions are placed at 20 Ggy^{-1} *(6)* which is approximately 25-30 % of the total MeBr production and constitutes 10-20% of the total anthropogenic emissions of MeBr to the atmosphere.

Pre-plant Soil Fumigations: The emission figures from pre-plant soil fumigations have been better defined in recent years although there are still uncertainties with respect to soil properties and application procedures in extrapolating the figures to the global scale (See Chapters 11-13). Emission estimates range widely depending mainly on the application procedure. For a tarped field the emission varies from 25 to 35% of the amount applied depending on the period of coverage *(12, 13)*, while for an untarped field the emission is as much as 90% of amount applied *(14)*. This source contributes between, 20-60 Ggy^{-1} of MeBr *(6)*. Chapter 10 describes some strategies for significantly reducing the emission source for MeBr, while Chapter 9 describes this as well for 1,3-dichloropropene.

Inter And Intra Phase Transport

Transport in Air: The major natural inputs of MeBr into the atmospheric compartment are oceanic and biomass burning while the major anthropogenic input is the emission from pre-plant soil fumigation. The major removal mechanisms are i) reaction with OH radicals and other chemical species, ii) photodissociation in the stratosphere, iii) flux into the oceans and, iv) uptake by soils and/or plants.

Transport In Water: The solubility of MeBr in water is between 16 and 18 g/L at 20°C. In soil water it is partially hydrolyzed to bromide ion. After fumigation the soil may be leached with water to prevent uptake of bromide ions by plants that may be planted after fumigation. It has been found that methyl bromide is able to diffuse through and be adsorbed by certain plastics (e.g., polyethylene). Thus drinking water pipes in the vicinity of fumigated fields could become contaminated within a few days of the use of MeBr *(11)*.

Transport In Soil: MeBr vapor has a density of 3.974 g/L at 20°C and thus is heavier than air. When injected into soil MeBr diffuses through the soil either by mass flow or molecular diffusion to depths of 60-240 cm *(15)*. Some MeBr hydrolyzes and some gets decomposed by microorganisms, but a major portion eventually dissipates into the atmosphere. The rate of MeBr degradation in soil is about 6-14% d^{-1} at 20°C *(16)*, and MeBr can be detected up to three weeks after fumigation, with the highest content being found in the upper 40 cm of soil layers and traces detectable to a depth of 80 cm *(17)*. Reversible sink processes such as sorption and dissolution, and irreversible sink processes such as reaction with the soil organic matter and hydrolysis that occur simultaneously during transport also affect the transport properties of the compound *(15)*. MeBr, in addition to being physically bound to the organic matter, is believed to methylate the carboxylic groups and the N- and S- containing groups of aminoacids and proteins in soil organic matter *(18)*. The sink (reversible and irreversible) capacity of a soil depends on its moisture and organic matter content. Chisholm and Koblitsky *(19)* found the sink capacities decreasing in the sequence peat, clay, and sand. In soils with high humic content the halflife of MeBr was 10 days, while in a less humic soil, it was 30 days, and in sand, about 100 days *(11)*. Reible *(20)* predicted that when soil organic carbon content was increased from 2 to 4% the MeBr emission rate would decrease from 45 to 37% following a tarped - 2 day, 25 cm deep - application. Gan et. al.*(21)* reaffirm the fact that soil organic mater content and moisture increase the sink capacities and decrease volatilization/emissions (See Chapters 10 and 11). In shallow top soils the major portion of MeBr dissipates into the atmosphere while simultaneously undergoing degradation as mentioned above. It is relatively persistent in the underlying strata where diffusion into the atmosphere is less likely. In such a situation, if low temperatures prevail, the water table is high, and the matrix is composed of a low density matter, then the potential for ground water contamination by MeBr is fairly high.

The irreversible sink processes produce significant amounts of bromide ion, which decreases to pre-fumigation levels in about one year. The highly mobile bromide ion is available for uptake by plants or can be leached by water *(22)* and contaminate groundwater.

Transformation / Sinks
MeBr undergoes transformations in various compartments of the environment (See Chapters 4-6). Most transformations result in the release of water soluble bromide anion while photolytic dissociation and reaction with hydroxyl radicals in the atmosphere yield bromine species, some of which later react with ozone and cause ozone depletion.
Geosphere
Hydrolysis: Methyl bromide hydrolyzes at neutral pH to methanol, bromide and hydrogen ion:
$$CH_3Br + H_2O \rightarrow CH_3OH + H^+ + Br^-$$
Hydrolysis rates under environmental conditions depend mainly on temperature and thus the half-lives vary from several hours to several days. The hydrolysis rates and half lives of MeBr at different temperatures are tabulated in Table 1. Recently Gan et al *(21)* have observed a similar dependence of the rate of hydrolysis on temperature. Moisture, as described earlier, enhances the sink/hydrolysis capacities of the soil.

Table 1. Hydrolysis rate constant (k) and half life of methyl bromide in water at different temperatures *(23)*

Temperature (°C)	Observed Rate Constant (s^{-1})	Half Life
17	1.07×10^{-7}	75.0 days
25	4.09×10^{-7} *(24)*	20.0 days
25	3.57×10^{-7}	21.3 days
35.7	1.65×10^{-6}	4.9 days
46.3	6.71×10^{-6}	1.2 days
100	1.28×10^{-3}	0.6 h

Hydrolysis is the primary route of degradation of MeBr in soils with a very low organic matter content. The adsorption isotherms in these soils were found to be linear but slopes were greatly reduced as moisture content increased *(25)*.

In soils containing organic matter, adsorption constitutes a reversible sink and is the primary process observed but reaction with organic matter, which constitutes the irreversible sink, is equally important. Methylation of carboxylic groups, on moist H-substituted peat, or of N- and S- containing groups is the predominant mechanism of this soil sink *(18)*.

Photolysis: The UV absorption cross section for MeBr of 174-262 nm with maximum absorption at 202 nm has been confirmed by many authors *(26-28)*. This is far below the shortest wavelength radiation reaching the earth's surface from the sun. Also, the photoactivation of methyl bromide, as well as its hydrolysis products, in water or in soil surfaces will differ from the gas phase activities of these processes. The study of the role of sunlight on hydrolysis under laboratory conditions showed very little effect *(29)*. All indications are that photolysis is not a significant fate process in condensed phases.

Microbial Degradation: In aqueous solution MeBr can undergo a variety of nucleophilic substitution reactions to yield methanol, methanethiol (MeSH) or dimethylsulfide (DMS). In anoxic environments having free HS^{-1}, MeBr will react to form MeSH and DMS which are subsequently attacked by methanogenic and/or sulfate-reducing bacteria *(30)*. Under aerobic conditions MeBr, acts as a methane analogue and MeBr is oxidized by cell suspensions of *M.capsulatus*. High MeBr levels (10,000 ppm) inhibit methane oxidation in soils but lower levels (1,000 ppm) of MeBr could be consumed by soil methanotrophs *(30)*. Some nitrifiers, such as *Nitrosomonas europea* and *Nitrosolobus multiformis*, and *Nitrococcus oceanus* could also be involved in this degradation. When provided with sufficient amounts of ammonia, *N europea* was able to degrade 98% of MeBr *(31)*. Experiments with ^{14}C-MeBr during fumigation events showed that about 10% of MeBr injected into strawberry fields was oxidized to [14]CO_2 *(32)*. Recently, Shorter et al. *(4)* reported soil microbial uptake of near ambient levels (pptv) of MeBr from the atmosphere. Uptake by natural microbial systems at near in-situ levels (picomolar) of MeBr in the oceans has also been observed. Unfiltered tropical ocean water samples removed MeBr about 40% more rapidly than filtered samples in 150 h experiments *(33)*. Indications are that MeBr uptake by microbes in both the geosphere and hydrosphere affords a major sink. Quantification of the size of the sink and of the amount taken up remains to be done.

Atmosphere

Photolysis In Air: In the upper stratosphere, above 25 km., photodissociation of MeBr is the dominant loss mechanism. Below this altitude, as less UV radiation is able to penetrate the atmosphere, the role of photolysis decreases. In the mid-stratosphere, between 20-25 km., photodissociation becomes competitive with loss by reaction with OH radical and diffusion. In and below the lower stratosphere, below 20 km., down to the troposphere, photodissociation becomes negligible and losses by diffusion and reaction with OH are of almost equal importance *(34)*. The end products of photodissociation of MeBr and reactions with hydroxyl radicals in the atmosphere are species such as Br, BrO, HBr, HOBr, BrCl and $BrONO_2$ *(2)*.

Atmospheric Sink - Reaction With OH radical: Reaction of MeBr with OH radical is the chief chemical removal pathway for MeBr from the lower troposphere. MeBr reacts slowly with hydroxyl radical:

$$CH_3Br + OH^\cdot \rightarrow CH_2Br^\cdot + H_2O$$

with a rate constant of about 3×10^{-14} cm^3/molecule/sec at 25°C *(35)*.

Laboratory data for the rate coefficients of MeBr *(36)* and methyl chloroform *(37)* reacting with hydroxyl radical, when combined with the estimated *(38, 39)* lifetime of the latter for removal by tropospheric OH deduced from measurements, resulted in an estimated OH removal lifetime for MeBr of about 2 (+/-0.5) yrs. *(2)*. This estimate has been lowered recently by 15% due to recalibration of atmospheric measurements of methyl chloroform *(40)* from which OH abundance was computed in the earlier estimate. Thus the estimated lifetime of atmospheric MeBr due to losses from reaction with OH is now about 1.7 (+/-0.2) yrs.

Other atmospheric removal processes, such as precipitation in the troposphere (estimated lifetime of 2,000 yrs.) and transport to the stratosphere followed by reaction with the OH radical and photodissociation (estimated lifetime of about 30-40 years), are additional sink processes for MeBr *(2)*.

Reaction With Stratospheric Ozone: In the stratosphere, UV radiation photodissociates MeBr and other brominated organic compounds to release Br atoms. Fig. 1 depicts the key bromine containing species in the atmosphere and shows the interconversion between reactive (Br and BrO) and reservoir (HBr, HOBr, BrCl, and BrONO$_2$) species *(2)*.

Yung et al.*(41)* proposed the following reaction for ozone loss due to reaction with halogens:

$$Br + O_3 \rightarrow BrO + O_2$$
$$Cl + O_3 \rightarrow ClO + O_2$$
$$BrO + ClO \rightarrow Br + Cl + O_2$$
$$\text{Net :} \qquad 2 O_3 \rightarrow 3 O_2$$

On a per molecule basis the efficiency of a bromine atom in destroying ozone is 30-60 times greater than that for a chlorine atom. This is due in part to the lower stability (and hence shorter lifetimes) of the bromine reservoir compounds. Such bromine-catalyzed ozone removal in the lower stratosphere has been thought to occur primarily via the reaction between BrO and ClO *(39)*. Thus the bromine induced ozone loss increases with increasing abundance of stratospheric chlorine. Bromine catalysis is most efficient in the lower stratosphere where the ozone concentration is largest.

Another possible catalytic cycle is that between BrO and HO$_2$ *(42)*.

$$Br + O_3 \rightarrow BrO + O_2$$
$$BrO + HO_2 \rightarrow HOBr + O_2$$
$$HOBr + h\nu \rightarrow OH + Br$$
$$OH + O_3 \rightarrow HO_2 + O_2$$
$$\text{Net :} \qquad 2O_3 \rightarrow 3O_2$$

The reaction rates for this reaction mechanism and others involving BrONO$_2$ or HOBr are still uncertain. If the rate of this reaction is faster as suggested by Poulet et al. *(42)*, then the ODP will increase.

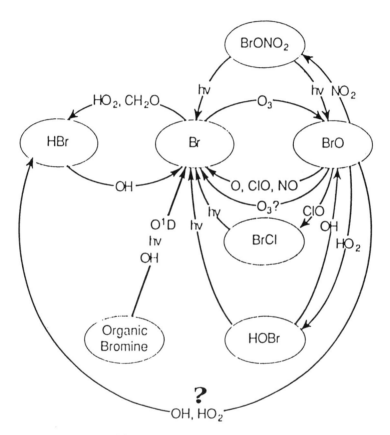

Fig. 1. Gas phase bromine cycle *(2)*.

The partitioning of bromine between different species in the stratosphere is not well known at this time. Studies of the formation of the relatively unreactive HBr by other reactions such as BrO + OH and Br + HO_2 would help to better understand bromine speciation in the stratosphere, which would, in turn, help in determining the contribution of bromine/MeBr to the global ozone depletion problem. As the estimated rate of HBr formation increases, the calculated ODP of MeBr decreases.

Oceanic Sink - Invasion Into Oceans: It has been found recently that the oceans, in addition to being a major source, are also a significant sink for atmospheric MeBr(43). The WMO, 1995 (6) report places the partial lifetime of MeBr with respect to oceans at 3.7 yrs. (1.5 to 10 yrs.). A very recent estimate by scientists at NOAA places the partial lifetime at 2.7 yrs. (2.4-6.5 yrs.) (44). NOAA scientists found 80% of the oceans to be undersaturated in MeBr representing a net annual sink of 8-22 Ggy^{-1}. The ocean is both a source and a sink everywhere. At regions of undersaturation the sink exceeds the source and vice versa. The net flux (source minus sink) varies with the atmospheric mixing ratio of MeBr (45). Recently Pilinis et al. (46) and Anbar et al. (47) have suggested by use of models that the oceans are probably a net source of MeBr considering that aquatic degradation would increase poleward in both hemispheres with lowering water temperature while production, being mainly biological, would follow chlorophyll or be relatively constant leading to a seasonal source of MeBr from polar waters. However, an investigation of summertime polar waters of the Laborador Sea and southern ocean do not substantiate the model results (48).

Soil Sink - Deposition Onto Soils: New investigations show that soils act as significant sinks for MeBr. Apparently due to adsorption and rapid biological degradation, this process creates a 42(+/-32) Ggy^{-1} sink. The estimated partial lifetime with respect to soil as a sink is about 3.4 yrs. (4). Preliminary studies by Woodrow et al., (49) support the existence of this sink. Despite there currently being too much variation and too little knowledge on the subject to precisely factor in this sink in the estimation of atmospheric lifetime, these estimates when employed in the lifetime estimates (4, 45) lead to larger latitude in uncertainty of the lifetime of MeBr.

Plant Sink - Deposition Onto Plant Leaf: Plants may take up atmospheric MeBr through leaf surfaces and metabolize it inside the tissue. This sink is currently under active investigation (See Chapter 6).

Atmospheric Budget of Methyl Bromide
The budget is controlled by the magnitude of the natural and anthropogenic sources and by the atmospheric and surface removal processes or sinks. Given the inadequacy and uncertainties in data cited above, an accurate budget cannot yet be calculated. The observational data indicate that the globally averaged atmospheric burden of MeBr is between 9 and 13 pptv which is equivalent to a total atmospheric loading of 150-220 Gg. with an interhemispheric ratio (IHR) of 1.3 (2). With two confirmed sinks, the atmospheric OH removal and the oceanic invasive flux, the best estimate of lifetime of atmospheric MeBr is 1.3 (0.8-1.7) yrs., with an ODP of 0.6.

The estimation of overall atmospheric lifetime from partial lifetime values of 1.7 yrs. with OH, 2.7 yrs. with oceans, and 3.4 yrs. with soil, yields a value of 0.8 yr. for MeBr. This leads to an ODP estimate of 0.36 (4). With this overall lifetime of 0.8 yr., there should be an overall (sum of natural and anthropogenic) source strength of 190-280 Ggy^{-1}. This is not in agreement with the estimated total emissions from all sources which is in the range of 100-194 Ggy^{-1} (5). This implies the existence of either additional sources or larger than current emissions from the known sources. Further, a 42 Ggy^{-1} uptake by soils is inconsistent with the observed IHR (=1.3) which is considered to be a reliable measurement. A larger loss to soil than previously estimated should lead to a decreased IHR (<1.3). If the new estimate of the soil uptake is firm, then there must be additional natural and/or anthropogenic sources of MeBr in the northern hemisphere to balance this sink and maintain the IHR at 1.3. This would then call for a reinterpretation of sources in the northern hemisphere.

Alternatively, the best estimate of the lifetime of MeBr, considering only the oceanic sink and the atmospheric OH removal processes, is 0.8 yr. (44) with an associated ODP between 0.45 and 0.49 (8). However, the WMO 1995 (6) report finds the overall lifetime of less than 0.6 yr. and ODP of less than 0.3 to be highly unlikely because of the constraints imposed by the observed IHR and total known emissions. The current best estimate of ODP of 0.45 for purposes of risk management is a reasonably conservative choice.

Human Exposure

In addition to stratospheric ozone depletion, risks due to human exposure are cited as reasons for considering severe limitations or an outright ban on use of MeBr. Some nations, such as the Netherlands, have imposed a ban on MeBr because of fatalities associated with the use of the chemical in greenhouses - a major agricultural use in that nation. The United States is concerned with both applicators and incidental exposure to MeBr vapors associated with uses in fumigation and particularly in open field soil fumigation operations.

Applicator field exposure includes operators of the application equipment - generally large tractors which carry methyl bromide cylinders, inject the liquid via shanks below the soil surface, and most often roll out a polyethylene tarp over the fumigated soil. Exposure potential also exists for field workers who seal the tarps at the end of the rows by shoveling dirt over them, and those who slit and then remove the tarps after the fumigation operation is complete, generally five days or so after treatment. Exposures are minimized by the following steps:

1. Chloropicrin is required to be added as a warning odorant if it is not an active ingredient in the fumigant.
2. MeBr is a restricted material which can only be applied by trained, certified applicators.
3. Field fumigation is accomplished quickly, so that only short term and sporadic exposure occur for individual personnel.
4. Respirators are an option for some situations.
5. Strict reentry rules are in effect for freshly fumigated fields.

Generally, these steps have worked effectively so that there are few if any field exposure situations with medical outcomes. This is in contrast to structural pest control in homes where a few fatalities have been reported.

The more substantial exposure issue is with downwind residents. In California MeBr is regarded as a reproductive toxicant and thus comes under the state's Proposition 65 requirements. This has heightened public awareness of use of MeBr, particularly in populated areas where strawberry fields may be interspersed among subdivisions and recreational areas. Also as a reproductive toxicant, MeBr is subject to toxicological data requirements of California's SB 951, the Birth Defect Prevention Act. Data requirements were imposed that called for submission of chronic toxicity data by March 1, 1996. Cancellation of MeBr was to occur if the data were not submitted by that date. Because the tests were not completed by the required date, a special session of the California legislature was convened by Governor Wilson, resulting in an extension of the deadline by two years.

The downwind exposure issue has been vigorously addressed by California's Department of Pesticide Regulation. A regulatory 24 hr. exposure limit of 816 $\mu g/m^3$ was set based upon assessment of current toxicological data. Emission estimates are made of specific fields, based primarily upon acerage and treatment rates. From these emission estimates, combined with the use of air dispersion modeling, buffer zones are then set for the minimum distance which must exist between the edge of the treated field and the nearest downwind residence. This approach is much like that being used to limit telone exposure in California (See Chapter 3). This approach to safeguarding public health offers the potential for allowing continued use of methyl bromide in California, at least for most situations in which it is needed. Chapter 13 describes how dispersion modeling may be used to determine downwind buffer zones from field fumigations. Chapter 14 describes an application of the methodology for chamber and warehouse fumigation and resulting MeBr downwind residues. Chapter 16 describes an *in situ* FTIR analytical technique useful for monitoring MeBr in air from structural and commodity fumigation operations.

Conclusions

New information on the sources, sinks, transport and transformation of MeBr has been developed in the past 3-4 years. This new information has changed the views of its behavior and resulted in a substantial lowering of ODP. Some are now beginning to question if MeBr is really as detrimental to the environment as it was made out to be from the earlier incomplete information. A threat from MeBr to the environment may exist, but may not warrant an outright ban. Like all synthetic chemicals MeBr has associated with it a certain risk but one which may be manageable. Better application technology, for example, might contain MeBr in the soil longer, so that chemical hydrolysis and microbial degradation become more competitive with volatilization. And the MeBr used to fumigate harvested commodities might be recovered and recycled, rather than ventilated to the atmosphere. These two areas alone represent possibilities for managing the use of MeBr, under carefully controlled conditions, so that the benefits of this chemical can continue to be realized, at least until economically viable alternatives are available.

Similar situations hold for other fumigants, such as telone (Chapters 3, 7, 9, 17), chloropicrin (Chapter 8), methyl isothiocyanate, phosgene etc. Common to all is the need for quantitative data on emissions and exposure so that good decisions can be made for both human health and agriculture.

References

1. Methyl Bromide Global Coalition (MBGC) **1994**: *"Annual Production and Sales for the Year 1984-1992"*. **June 4' 1994**, Washington D. C.
2. Methyl Bromide Global Coalition (MBGC) **1992**, *The Methyl Bromide Science Workshop Proceedings*. Edited. by M. K. W. Ko and N. D. Sze. Arlington, Virginia: Atmospheric and Environmental Research, Inc., Cambridge Massachusetts.
3. Ragsdale, N. N. and Wheeler, W. B., *Rev Pestic Toxicol* **1995** *3* 21
4. Shorter, J. H., Kolb, C. E., Crill, P. M., Kerwin, R. A., Talbot, R. W., Hines, M. E., and Harriss, R. C. *Nature* **1995**. *377(10)*. 717,.
5. Butler, J. H., *Atmospheric Environment,.* **1996** *30(7)*., i.
6. WMO **1995**: *Global Ozone Research and Monitoring Project - Report # 37., Scientific Assessment of Ozone Depletion 1994*. World Meteorological Organization, Geneva 1995.
7. Andreae, M. O. et. al., *J. Geophys. Res.*, in press **1996**
8. Khalil, M. A. K., Rasmussen and R. Gunawardena *J. Geophys Res. -D2* **1993** *98*, 2887.
9. Singh, H. B. and M. Kanakidou, *Geophys. Res. Lett.,* **1993** *20*(2), 13.
10. NOAA *"Bromine Latitudinal Air/Sea Transect 1994"*. Technical Memorandum, National Oceaninc and Atmospheric Administration, ERL CMDL - 10, Boulder CO 1996:
11. EHC 166: *Environmental Health Criteria 166 - Methyl Bromide* International Programme on Chemica Safety. Pub. by: World Health Organization 1995
12. Yagi, K., Williams, J., N-Y. Wang, R. J. Cicerone., *Science*, **1995** *267,* 1979.
13. Seiber, J. N., Woodrow, J. E., Honaganahalli, P. S., LeNoir, J. S., and Dowling K. C., *Proceeding of 1995 ACS Symposium on Fumigants - Chicago, 1996*.
14. Majewski, M. S., McChesney, M. M., Woodrow, J. E., Prueger, J. H., and Seiber, J. N., *J. Environmental Quality*, **1995** *24(4)*, 742.
15. Brown, D. and Rolston, D. E., *Soil Science* **1980**, *130,*. 68.
16. Daelemans *"Uptake of methyl bromide by plants and uptake of bromide from decontaminated soils"* Leuwen, Belgium, Catholic University, Dissertation, **1978**
17. Lepschy, J., Stark, H. and Sub, A., *Gartenbauwissenschaften*, **1979** *44*, 84,.
18. Maw, G. A. and Kempton R. J. "Methyl bromide as a soil fumigant." Soils Fertil. **1973**. *36*, 41.
19. Chisholm, R. D. and Koblitsky, L., *J. Econ. Entomol.*, **1943**, *36*.,. 549.
20. Reible, D. D. *J. Hazard Mater.* **1994** *37*, 431,
21. Gan, J., Yates, S., Wang, D. and Spencer, W. F., *Environ. Sci. Technol.* **1996** *20*, 1629.
22. CEC. *Report of the Scientific Committee for Pesticides on the use of methyl bromide as a fumigant for plant culture media*. Luxembourg, Commission of the European Communities, 15-33 (EUR 10211) 1985.

23. Moelwyn-Hughes, E. A. *Proc. R. Soc*, **1938**, *A164* 295.
24. Mabey, W. and Mill, T. *J. Phys Chem Ref. Data*, **1978**, *7(2)*, 383.
25. Arvieu, J. C. *Acta Hortic.* **1983**, *152*, 267.
26. Robbins, D. E. *Geophys Res Lett*, **1976** *3*, 757.
27. Uthman, A. P., Demlein, P. J., Allston, T. D., Withiam, M. C., McClements, M. J., and Takacs, G. A. *J. Phys Chem*, **1978** *82*, 2252.
28. Molina, L. T., Molina, M. J. and Rowland, F. S., *J Phys Chem*, **1982** *86*, 2672.
29. Gentile, I. A., Ferraris, L., Crespi, S. and Belligno, A. *Pestic. Sci,.* **1989** *25*. 261.
30. Oremland, R. S., Miller, L. G. and Stormier, F. E., *Env. Sci and Technol*, **1994**, *28*, 514
31. Rasche E., Hyman, M. R. and Arp, D. J., *Appl. Environ. Microbiol.*, **1990**, *56*,. 2568,.
32. Miller L, T. Connel, and R. Oremland et al., unpublished .data
33. King, D. B., Pilinis, C. and Saltzman, E. S. *EOS Trans. AGU* **1996** *76 (3)*, OS7.
34. Robbins, D. E. *Geophys. Res. Let,.* **1976**, *3*, 213.
35. NASA *Chemical kinetics and photochemical data for use in stratospheric modeling.* Evaluation No. 10: NASA panel for data evaluation. Pasadena California Institute of Toxicology, Jet Propulsion Laboratory, National Aeronautics and Space Administration.
36. Mellouki, A., R. K. Talukdar, A. Schmoltner, T. Gierczak, M. J. Mills, S. Solomon, and A. R. Ravishankara, *Geophys. Res. Lett.* **1992**, *19(20)*, 2059
37. Talukdar, R. K., Mellouki, K. A., Schmoltner, A. M., Watson, T., Montzka, S. and Ravishankara, A. R. *Science* **1992**/ *MBSW* **1992**.
38. Prinn, R. G., Cunnold, D., Simmonds, P., Alyea, F., Boldi, R., Crawford, A., Fraser, P., Gutzler, D., Hartley, D., Rosen, R. and Rasmussen, R. *J. Geophys. Res.*, **1992**, *97*, 2445.
39. WMO **1992**:, *Global Ozone Research and Monitoring Project - Report # 25.*, *Scientific Assessment of Ozone Depletion-1991.* World Meteorological Organization, Geneva 1991.
40. Prinn, R. G. et al., *Science.* **1995**, *269(187)*.
41. Yung, Y. L., Pinto, J. P., Watson, R. T. and Sander, S. P., *J. Atmos Sci,* **1980**, *37*, 339.
42. Poulet, G., Pirre, M., Maguin, F., Ramaroson, R. and LeBras, G *Geophys. Res. Lett.* **1992**/ *MBSW*, **1992**.
43. Lobert, J. M., Butler, J. H., Montzka, S. A., Geller, L. S., Mayers, R. C. and Elkins, J. W. .*Science*, **1995** *267*, 1002.
44. Yvon, S. A. and J. H. Butler, *J. Geopphys. Res. Lett.*, **1996**, *23*, 53.
45. Butler, J.H., *Geophys. Res. Lett.*, **1994**, *21(3)*, 185
46. Pilinis, C., King, D. B., and Saltzman, E. S. *J. Geopphys. Res. Lett.*, **1996**, *23*, 817.
47. Anabar, A. D., Yung, Y. L. and Chavez, F. P.*J. Global Biogeochemical Cycles* **1996** *10*, 175.
48. Moore, R. **1996** Unpublished data. University of Dalhousie, Halifax, Nova Scotia, Canada.
49. Woodrow, J. E., J. LeNoir and J. N. Seiber **1996**, Unpublished data.

Chapter 2

Role of Soil Fumigants in Florida Agriculture

J. W. Noling

Institute of Food and Agricultural Sciences, Citrus Research and Education Center, University of Florida, 700 Experiment Station Road, Lake Alfred, FL 33850

Historically, preplant soil fumigants have had a very global, profound, and stabilizing influence on production agriculture, and catalyzed the development of several high value, multiple cropping systems, In some cases, fumigants have been adapted almost to the exclusion of all other soil pest management strategies because of their superior broad-spectrum control efficacy and consistent enhancement of crop growth, development, yield, and quality. In general, soil fumigation has allowed growers to repeatedly use the same fields for crop production each year, 2) to specialize in a few crops and integrate crop production cycles with market requirements, 3) to take increased advantage of financial investments in property and land improvements, and 4) to minimize capital investments in farm machinery and labor requirements. Certain agricultural industries will be adversely affected by the removal of soil fumigants from commercial use unless new, economically and environmentally acceptable integrated pest management (IPM) strategies are developed and implemented.

Prior to 1950, Florida vegetable culture can best be described as nomadic. One or two vegetable crops were produced in sequence on rented land after expensive clearing operations had been performed or after long pasture rotations to minimize soil borne pest and disease problems (1,2). For example, tomato farmers found it profitable to clear, ditch, dike, install pumps and wells and construct graded roads to virgin land each season in order to escape soil borne pest and disease problems on previously cropped soils (3). Once a problem developed, Florida truck farmers (as they were called at the time) were then forced to migrate from one field or area to another, opening new land and abandoning the old to avoid the crop pests which, perforce, became more severe with re-use of the same fields. As urban growth

0097–6156/96/0652–0014$15.00/0
© 1996 American Chemical Society

increased, suitable land became more difficult to locate as well as prohibitively expensive, both in terms of purchase or leasing, and land preparation (2,4). Due to these constraints, Florida vegetable farmers increasingly adopted the use of soil fumigants to managed established weed, nematode, and disease pests within their fields(5-12). Reverting back to such a system would no longer be considered a viable option due to widespread urbanization and environmental and water management regulatory policies (13,14).

Historical Development and Use of Fumigants

Discovered in 1869, the first fumigant material to become available for agricultural use in Florida was carbon bisulfide (15). Because of its flammability, high rates of application, cost, and pest control inconsistency, carbon disulfide was seldom recommended or extensively used within Florida (15-17). Only recently, has interest been renewed for the use of carbon bisulfide (18). In 1920, Russell (19) reported excellent control of the root-knot nematode with chloropicrin which was later confirmed by other workers (16,20). It was demonstrated that treating soil with chloropicrin cured the problems of "soil sickness" and restored the soil to a higher productivity than could be obtained from years of crop rotation or fallow. In the United States, the war surplus of chlorpicrin was shipped to Hawaii, where beginning in 1927, it was applied as a preplant fumigant to pineapple (6,15). Its use in Hawaii continued until the supply was exhausted, after which very little chloropicrin was used for agricultural purposes.

In 1935, Neller and Allision (21) discussed the use of a machine for subsurface treatments of nematode infested soil with chloropicrin and carbon bisulfide. After considerable delay since its discovery, chloropicrin was commercially introduced as a soil fumigant to Florida in 1937 (22). It was first used on a limited basis by nurserymen and greenhouse operators as a substitute for soil steaming. In 1941, Taylor and McBeth (23) proposed the spot treatment method of fumigation wherein chloropicrin was used to fumigate small seedbed areas within fields. Because of the high material cost, very little development of large scale application equipment, other than hand gun type injectors, had been undertaken in Florida (24).

It was not until D-D (1,3-dichloropropene, 1,2,-dichloropropane and related C3 hydrocarbons) was developed in 1943 (25) and repeatedly demonstrated to be an effective nematicide, was any concentrated effort made on developing specialized field application equipment. By 1945, Shell Chemical Corporation had developed a tractor drawn cart with applicator, shanks, pumps, and tank for soil fumigation and offered the first custom application service. As a result, D-D, the first really effective and inexpensive nematicide for general field use, was commercially introduced into Florida in 1945 (22,26).

In 1945, Christie (27) reported that ethylene dibromide (EDB) gave excellent control of root-knot nematode (Meloidogyne spp.) in preliminary tests. The Dow

Chemical Company was also testing EDB and soon introduced it as a low cost fumigant under the trade name Dowfume W. Other companies were soon marketing EDB under various trade names. Within a few years of 1945, use of D-D and EDB increased significantly because of the oftentimes 'astonishing' improvements in plant growth and yield which resulted following their use (22,26). The most significant outcome from these early soil fumigation trials was the demonstration of the very great importance of plant-parasitic nematodes in regulating crop productivity (28). Because use of these soil fumigants made the difference between an excellent yield and no yield, many growers began adopting soil fumigation as a general practice (1,7,12,22,29).

Methyl bromide was first reported to be effective as a soil fumigant in 1940 (30) and 1942 (16). As a soil fumigant, it was applied under a gas tight cover (31), with the gas introduced by special applicator into the space between the cover and soil surface (raised tarp method). At the time however, the use of methyl bromide was only considered warranted for small plots such as greenhouse beds or benches in which it was identified as a replacement for steam sterilization. In 1953, Wilhelm et al. (32) reported the synergistic effect of methyl bromide-chloropicrin mixtures on control of Verticillium wilt of strawberries. Extensive field use of methyl bromide in Florida would not come until some 10 years later.

Much of the experimental soil fumigation work in Florida dates back to the spring of 1945 (33). In these studies, chloropicrin and D-D significantly increased crop yields while seedbed studies showed that plots fumigated with methyl bromide yielded significantly more plants suitable for transplanting than nonfumigated plots. The specialized equipment required for field application was expensive, and because of its high cost, was only used by large Florida growers. By 1949 soil fumigation with D-D and EDB had expanded to such an extent in many areas of the west and southeast, that the first mass production of field application equipment was undertaken (24).

At the time however, it was also apparent from field observations, that plant stunting and inconsistent responses in crop growth and yield could be attributed to physical and edaphic factors as well as to the detrimental impacts of some of these fumigants on soil nitrifying organisms (33). Soil fumigation efficacy was shown to be affected by physical soil factors such as soil texture, pH, organic matter content, temperature, and moisture (34-37), formulation (9) and application methods (11,15,16,38-40). It was also shown that use of EBD did not interfere with conversion of complex organic and ammoniacal forms of nitrogen, but methyl bromide did significantly depress nitrification (41). It was soon discovered however, that the low nitrate-nitrogen values and high ammoniacal nitrogen levels following soil fumigation could be alleviated without phytotoxic plant responses if nitrate forms of fertilizer were used (2).

It was also common knowledge within these early years of fumigant use, that no fumigant, regardless of method or rate of application, could be effective in preventing the development of root-knot nematode on a susceptible crop that remained in the field for an extended period, i.e. 3 months (*33*). The general rule was: The longer the life of the plant, the greater the need for protection against infestation (*42*). Because of these shortcomings, and even though extensively used in some cases, none of the fumigants were vigorous encouraged by cooperative extension personnel for field use except as a last resort (*33*).

Fumigation for vegetable production began in earnest during the early 1950's with in-furrow applications of nematicides to control root-knot nematode and other nematodes which, as indicated, quickly became an economic problem in re-cropped fields. In 1954, DBCP (1,2-dibromo-3-chloropropane; Nemagon) was determined to be a highly effective nematicide (*43*). Sodium methyldithiocarbamate (Vapam) was introduced circa 1956 (*44*) and methyl isothiocyanate mixed with D-D and released as DD-MENCS (Vorlex) about 1960 (*45*). This in-the-row, open bed (without plastic mulch covering) soil fumigation practice was at the time superior to no treatment but proved to be inadequate for protecting late maturing, long-term crops or for crops grown in fields heavily infested with disease organisms (*22*). Because of this, these early broad spectrum fumigants were used primarily in seedbed production until the appearance of vascular wilt diseases occurred (IE. Fusarium wilt race 2 & 3) in the sandy soils of the vegetable producing areas of the state.

As of 1953, the majority of Florida tomatoes were still grown on newly cleared sandy land, the tomato grower cooperating "with the cattleman to the benefit of both" (*42*). The list of yield limiting soil-borne tomato diseases which became important after 1 or 2 crop cycles included Fusarium wilt, Rhizoctonia, Southern blight, black-spot, early blight, buckeye-rot, bacterial wilt, root-knot, and other diseases due to nematodes. At the time and still relevant today is the fact that there are no commercially available crop cultivars resistant to all races or variants of these diseases. By 1960, the scarcity of 'new land' for vegetable crop production had encouraged growers to returned to fields converted in the past to pastures when diseases, weeds, and nematodes complicated vegetable production (*46*).

Broadcast soil fumigation had been too expensive on a field scale to be practical for wide row crops such as tomatoes (*41*). However, in-the-row soil fumigation (*29*), as practiced since 1949 on short-term crops for nematode control, had proved inadequate for the long-term trellised tomatoes gown for pink harvest and for crops grown in fields heavily infested with disease organisms (*41*).

In 1962, Geraldson (*2*) reported that a plastic mulch bed covering proved effective for weed control, significantly reduced fertilizer leaching losses, and restricted soil moisture fluctuations in vegetable crops. This study was the first to justify the benefits with the high cost of mulch and the cost of removal.

By the mid 1960's, soil fumigation had become a common practice for controlling nematodes in some crops. For example, the use of nematicides in strawberry production was such an established practice in Florida that 97 percent of strawberry acreage was treated with a soil fumigant in 1964 (*47*). Within the previous seven years (1957-64), the use of polyethylene mulch and herbicides in conjunction with the pre-plant fumigants had resulted in a four-fold increase in average strawberry yields (48). However, even at this time, these growing practices (use of these fumigants) were not considered sufficient for control of damping-off caused by Rhizoctonia solani. It was for this specific shortcoming that continued research with new fumigants such as methyl bromide in combination with chloropicrin were initiated (47).

In 1965 an integrated systems approach to soil pest management for sandy soils was introduced to the Florida vegetable industry to solve the "old land" problems which developed in repeatedly cultivated fields. This new systems approach to pest management allowed and encouraged vegetable growers to use the same fields for vegetable crop production each year, taking advantage of their financial investment in drainage and irrigation systems and site location relative to climate. The system which was advocated combined broad spectrum soil fumigation using methyl bromide and chloropicrin with full-bed plastic film mulch, and a constant water table utilizing seep irrigation to provide a minimum stress environment in the root zone of the crop. By 1977, sixty-four percent of the tomato acreage in peninsular Florida was fumigated with formulations of methyl bromide-chloropicrin, a 15% increase over the preceding 4 year period (1).

Currently, soil fumigation with methyl bromide chloropicrin for tomatoes occurring on effectively 100% of the acreage, begins in early July and continues through February of each year. Two formulations, 98% methyl bromide and 2% chloropicrin and 67% methyl bromide and 33% chloropicrin, are most extensively used with in-row application rates of 168-224 kg/ha. It is delivered under a full-bed plastic mulch (i.e., a 75-80 cm wide raised bed covered the entire width with 0.0025 cm thick polyethylene plastic film). Although single stream fumigation resulted in an increase in yield (38) it has been the practice, especially with the development of Fusarium wilt, incited by Fusarium oxysporum Schl. f.sp. lycopersici in Florida to use multiple streams, injected with 3 chisels 20-25 cm apart and 15-20 cm deep to protect the tomato crop from soil-borne pathogens, nematodes, and weeds. Fertilizers are placed on the top or shoulders of the bed. Subsequent development of disease resistant cultivars and containerized transplants has further refined the integrated system for added crop protection from soil pests.

Development and use of the polyethylene mulch, high analysis soluble fertilizers, seep irrigation, and methyl bromide chloropicrin soil fumigation has allowed and assured high quality vegetable crop yields on lands of low natural fertility infested with nematodes, soil-borne disease organisms and numerous weed pests. Protection of plants from moisture and nutrient stress, competition by weeds and the root pruning associated with cultivation apparently contributed to increased tolerance of the tomato plants to root-knot nematode infection (49) and depressed expression of symptoms of southern blight of tomato (50). Polyethylene mulch has become an integral component of the system and has been advocated for numerous reasons: 1) to preserve and protect raised soil beds against erosion, 2) to retard leaching of nutrients, 3) to retard soil evaporation and to maintain consistent and nearly optimum soil moisture, 4) to improve the efficacy of in-the-row fumigation with methyl bromide/ chloropicrin, 5) to retard re-infestation of fumigated soil, 6) to prolong weed control, 7) to reduce damage of crop plant roots associated with cultivation, and combined with the high water table, 8) to limit plant roots to the fumigated soil volume which protects the plant longer from root-infesting disease organisms (*49*).

Since 1960, many different chemicals capable of killing some of the specific pests attacking vegetables in Florida have been field evaluated. Six soil fumigant have been used at one time or another for tomatoes: including methyl bromide-chloropicrin; DD-MENCS [chlorinated C3 hydrocarbons (80% D-D) + 20% methyl isothiocyanate] (Vorlex); sodium methyl dithiocarbamate (Vapam); 1,3-D (1,3-dichloropropenes (D-D, Telone II); EDB (ethylene dibromide) (Dowfume, Soilfume, W85); and DBCP (1,2-dibromo-3-chloropropane (Nemagon, Fumazone). In general methyl bromide chloropicrin (67/33) and DD-MENCS shared the major role in fumigation of about 70% of the tomato fields. During the period of 1974 to 1978, use of methyl bromide and chloropicrin increased from 30-43% of fields, while DD-MENCS decreased from 39 to 30% (*1*).

Of these fumigants, some have shown promise for certain pests (primarily nematode) under Florida conditions (*51-54*), but none have proven to be the ideal material, that is, one which possesses herbicidal, fungicidal, and nematicidal properties and, at the same time, did not leave toxic residues in the soil for an inconvenient length of time or pose potentially serious environmental risks to surface or groundwater supplies. For example, in May 1979, DBCP was discovered in groundwater in California, and shortly thereafter withdrawn from most of the global marketplace (*15*). In 1983, EDB was similarly banned for environmental contamination.

The benefits of methyl bromide soil fumigation have not been confined exclusively to soilborne pest and disease control within Florida vegetable cropping systems. For example, a 38% corn and grain sorghum yield reduction which occurred during the period of 1977-1981 was reversed by soil fumigation with methyl bromide-chloropicrin (*55*). In some cases the large increase in crop yields associated with methyl bromide fumigation is attributed to weed control (*56*).

None of the nonfumigant nematicides have proven to be consistently effective against root-knot nematodes as the fumigant type nematicides such as methyl bromide, while some have actually enhanced yellow nutsedge problems (*54*). Many of the nonfumigant nematicides have also been identified as chemical which are mobile in soil and have been found to contaminate groundwater as a result of agricultural use (*6*).

Role of Nematodes in Florida Crop Production

The primary nematode parasites of vegetable crops grown throughout the state of Florida include the root-knot nematodes (Meloidogyne spp.) and the sting nematode

(Belonolaimus longicaudatus), either of which may cause extensive root damage and yield loss. Other species of plant-parasitic nematodes are found in fine sandy and calcareous Rockdale soil in the southeastern part of the state. The root-knot nematode is a serious and ubiquitous pest of vegetables on both soil types. On sandy soils, sting, awl and stubby root nematode can affect vegetable crops but these species are rare or unknown on Rockdale soil. On the heavier calcareous Rockdale soils of southeastern Florida the reniform nematode is common and known to damage some vegetable crops (*54*).

With the exception of south Florida and small areas of muck land vegetable production, Florida vegetables are grown on fine sandy soils (95-97% sand, 3-4% silt, 0.2-0.5% clay) with low soil organic matter content (< 2%). In south Florida, vegetable soils are shallow, underlain by limestone, and contain 33-67% small limestone rock and gravel, mixed with fine sand or fine sandy loam soil. Much research data and field observations suggest that root-knot nematode causes greater damage on light sandy soils than on the heavier sandy loam or clay soils. Poor water and nutrient retention on the sandier soil undoubtedly contributes to poorer crop growth on the sandier soils.

Due to the extensive use of methyl bromide chlropicrin formulations as a broadspectrum fumigant nematicide in Florida, nematode induced crop loss is probably less than 1% of total production. Plant damage, when it is observed, generally occurs in areas where fumigants are not applied. These areas often are ends of rows where fumigants delivery is discontinued prematurely or in areas where exhausted fumigant cylinders are changed. Soil fumigants are also not generally applied in small farming operations of less than 5 acres where nematode damage is consequently often severe. Small farm (<5 acres) vegetable production, although important to the local economy, represents only a tiny fraction of total production and therefore has little influence on the magnitude of total vegetable crop loss in Florida.

Scenarios describing changes in crop production and pest control costs with loss of specific pesticide registrations have been previously estimated. In most analyses, the cost of plastic polyethylene mulch is not considered as part of the pest control costs because the mulch is a horticultural prerequisite for successful production irrespective of pests (*41,49*). With the loss of methyl bromide, other chemical means of nematode management such as 1,3-D dichloropropane (Telone), which is compatible with current production practices and available application equipment, would likely be adopted for broad spectrum soil pest control (*18,57*). At application rates which would provide similar levels of pest control, average production losses of 10-20% would likely occur in many producing areas of the state (*58*).

Other nonchemical pest control tactics have also been assessed. For example, crop rotation cannot in most cases be considered as an effective substitute for methyl bromide soil fumigation because of the wide host range of the root-knot nematode and because other major disease organisms are also not effectively controlled (*41*).

For example, some efforts have been made to rotate vegetable fields with small grains and pasture grasses to manage and prevent the buildup of soil borne pest and disease organisms. It was soon discovered however that the wide variety of nematode and soilborne disease pests which occur in Florida complicated the selection of rotatable crops, because crops which will reduce some species of nematodes and disease may favor the increase of others. Long term, minimal profit generating rotations are considered a major economic constraint (57). Commercially acceptable root-knot nematode resistant crop varieties which retain nematode resistance under high Florida temperature conditions are also not available and likewise cannot be seriously considered as an effective alternative for to methyl bromide soil fumigation.

It is also clear from previous research trials in Florida that soil fumigation with methyl bromide does not completely eradicate root-knot nematodes from soil. (59). However, root galls which do form on tomato roots in methyl bromide fumigated soil generally occur on secondary roots during late stages of crop growth with little influence on final crop yield. It is also clear that early season reduction of root-knot nematode is also frequently accompanied by a delay or overall decrease in wilt disease incidence. Late season population development of root-knot nematode becomes important when future crops within the cropping system are considered (57,60). Most vegetable crops in Florida are produced on old land repeatedly planted to nematode susceptible crops. In many areas a single spring or fall crop is produced per year and maintained weed fallow during the off-season. In some producing areas however, up to 80% of the fall crop acreage is double cropped', usually immediately followed by a cucurbit crop such as squash or melon. For example, watermelon yields may be reduced 50% when sequentially grown after tomato under the same plastic mulch (60). Realistically, crop loss estimation should consider the cropping system, including carry-over residual effects into subsequent crops (60-62).

The effect of root-knot nematode on vegetable crop yield results not only in a reduction in yield, but also in non-uniform maturation of the crop during periods of increasing weed competition. Plant senescence is accelerated and fruit generally mature earlier but non-uniformly. Numerous individual pickings may be required to remove harvestable fruit from the field which increases picking costs and lowers total crop net revenue. Conversely, methyl bromide treated crops generally have prolonged vegetative growth, with more uniform fruit maturity so that harvesting can usually be completed in 2-3 pickings.

Even though at cost to farmers, soil fumigation has become an integral part of economic crop production in Florida (60-62). The high costs of production and competition with foreign imports have demanded maximum production which at present only soil fumigation can provide. Additionally, preplant application of methyl bromide to soil controls a wider variety of pests at lower cost than combined use of many specific pesticides. In some instances, pesticides are not available to effectively manage a specific tomato pest problem, e.g., Fusarium oxysporum on tomato. Nonfumigant nematicides, which are expensive, have not proven to be consistently effective against root-knot nematode as indicated by the lower average yields and higher postharvest root gall ratings of field trials which have been performed since 1960 (*63,64*). In the event that methyl bromide or other soil fumigants were not available for use in the future, overall pesticide use should increase, with associated risks to field workers and environment. Crop loss and pest control costs should also be expected to increase (*57,60-62,66*). In this regard, alternative pest control practices including cultural, chemical, and biological which are more friendly to the user and environment, need to be developed prior to the elimination of current methods (*57,67*).

Literature Cited

1. Overman, A.J.; Martin, F.G. Proc. Fla. State Hort. Soc. **1978**, 91, 294-297.
2. Geraldson, C.M. Proc. Fla. State Hort. Soc. **1962**, 75, 253-260.
3. Winchester, J.A.; Hayslip, N.C. Proc. Fla. State Hort. Soc. **1960**, 73, 100-104.
4. McPherson, W.K. Proc. Soil and Crop Sci. Soc. Fla. **1965**, 25, 303-310.
5. Winchester, J.A.; Averre III, C.W. Proc. Soil Crop Sci. Soc. Fla. **1966**, 26, 31-34.
6. Johnson, A.W.; Feldmesser, J. **1987**. Nematicides-A historical review. Pp. 448-454 in J.A. Veech and D.W. Dickson, eds. Vistas on Nematology. Hyattsville, MD: Society of Nematologists. DeLeon Springs, FL: E.O. Painter Printing.
7. Johnson, H.; Paulus, A.H; Wilhelm, S. Calif. Agric. **1962**, 16, 4-6.
8. Jones, J.P.; Overman, A.J; Geraldson, C.M. Plant Dis. Reptr. **1971**, 55, 26-30.
9. Overman, A.J.; Jones, J.P. Proc. Fla. Hort. Soc. **1980**, 93, 248-250.
10. Overman, A.J.; Jones, J.P. Proc. Fla. State Hort Soc. **1984**, 97, 194-197.
11. Rhoades, H.; Holmes, E.S; Scudder, W.T. Proc. Fla. State Hort. Soc. **1962**, 75, 125-129.
12. Williamson, C.E.; Tammen, J.; Hannon, C.I; Denmark, C. Proc. Fla. State Hort. Soc. **1955**, 68, 370-373.
13. Bewick, T.A. Acta Hort. **1989**, 25, 61-72.
14. Reynolds, J.E.; Norberg, R.P. Proc. Soil and Crop Sci. Soc. Fla. **1983**, 42, 122-126.
15. Lembright, H. W. J. Nema. **1990**, 22, 632-644.
16. Taylor, A.L. Proc. Soil Sci. Soc. Fla. **1942**, 4, 126-140.
17. Watson, J.R. and C.C. Goff. **1937**. Control of root-knot in Florida. Fla. Agric. Experiment Station Bulletin #311, 22 pp.
18. United States Department of Agriculture. **1993**. Methyl bromide substitutes and alternatives: A research agenda for the 1990s. Washington D.C. 21-23 September 1992.
19. Russell, E.J. Royal Hort. Soc. J. **1920**, 45, 237-246.
20. Young, P.A. Phytopath. **1940**, 30, 860-865.
21. Neller, J.R.; Allison, R.V. Soil Sci. **1935**, 40, 173-178.

22. Perry, V.G. Proc. Soil Sci. Soc. Fla. ?. **1952**,
23. Taylor, A.L.; McBeth, C.W. Proc. Helm. Soc. Wash. **1941**, 8, 53-55.
24. Russell, J.C. Proc. Soil Sci. Soc. Fla. ?. **1952**,
25. Carter, W. Science **1943**, 97, 383-384.
26. Christie, J.R. **1959**. Plant nematodes. Their bionomics and control. Jacksonville, FL: L. and W.B. Drew.
27. Christie, J.R. Proc. Helm. Soc. Wash. **1945**, 12, 14-19.
28. Steiner, G. Proc. Soil Sci. Soc. Fla. **1942**, 4, 71-120.
29. Walter, J.M.; Delsheimer, E.G. Proc. Fla. State Hort. Soc. **1949**, 62, 122-126.
30. Taylor, A.L.; McBeth, C.W. Proc. Helminthol. Soc. Wash. **1940**, 7, 94-96.
31. Taylor, A.L. Proc. Helminth. Soc. Wash. **1941**, 8, 26-28.
32. Wilhelm, S.S.; Storkan, R.C.; Sagen. J.E. Phytopath. **1961**, 51, 744-748.
33. Spencer, E.L.; Burgis, D.S.; Jack, A. Proc. Soil Crop Soc. Fla. **1952**, 12, 72-75.
34. Goring, C.A.I. Adv. Pest Control Res. **1962**, 5, 47-84.
35. Goring, C.A.I. Ann. Rev. Phytopath. **1967**, 5, 285-318.
36. Jones, J.P.; Overman, A.J. Plant Dis. Reptr. **1976**, 60, 913-917.
37. Christie, J.R. Proc. Fla. State Hort. Soc. **1949**, 62, 117-118.
38. Overman, A.J.; Jones, J.P. Proc. Fla. Hort. Soc. **1977**, 90, 407-409.
39. Munnecke, D.E.; VanGundy, S.D. Ann. Rev. Phytopath. **1979**, 17, 405-429.
40. Rhoades, H.; Holmes, E.S; Scudder, W.T. Proc. Soil Sci. Soc. Fla. **1962**, 22, 147-152.
41. Overman, A.J.; Jones, J.P.; Geraldson, C.M. Proc. Fla. Hort. Soc. **1965**, 78, 136-142.
42. Walter, J.M. Proc. Soil Sci. Soc. Fla. **1953**, 13, 57-60.
43. McBeth, C.W. Plant Disease Reptr. **1954**, 227, 95-97.
44. Lear, B. Plant Dis. Reptr. **1956**, 40, 847-852.
45. Rhoades, H.L. Plant Dis. Reptr. **1961**, 45, 54-57.
46. Overman, A.J. Proc. Fla. State Hort. Soc. **1961**, 74, 201-204.
47. Overman, A.J. Proc. Soil Crop Sci. Soc. Fla. **1965**, 25, 351-356.
48. Overman, A.J. Proc. Fla. State Hort. Soc. **1963**, 76, 114-119.
49. Overman, A.J.; Jones, J.P. Proc. Soil Crop Soc. Fla. **1968**, 28, 258-262.
50. Jones, J.P.; Overman, A.J.; Geraldson, C.M. Phytopath. **1966**, 56, 929-932.
51. Jones, J.P.; Overman, A.J. Phytopath. **1971**, 61, 1415-1417.
52. Jones, J.P.; Overman, A.J.; Geraldson, C.M. Plant Dis. Reptr. **1972**, 56, 953-956.
53. Jones, J.P.; Overman, A.J. Plant Dis. Reptr. **1978**, 62, 451-455.
54. McSorley, R.; McMillan, R.T.; Parrado, J.L. Proc. Fla. Hort. Soc. **1985**, 98, 232-237.
55. Mislevy, P.; Overman, A.J.; Dantzman, C.L. Proc. Soil and Crop Sci. Soc. Fla. **1984**, 43, 142-145.

55. McSorley, R.; McMillan, R.T.; Parrado, J.L. Proc. Fla. Hort. Soc. **1986**, 99, 350-353.
57. Noling, J.W.; Becker, J.O. J. Nema. **1994**, 26, 573-586.
58. Noling, J.W.; Overman, A.J. J. Nema. **1989**, 21, 577.
59. Perry, V.G. Proc. Fla. State Hort. Soc. **1953**, 66, 112-114.
60. Spreen, T.H.; VanSickle, J.J.; Moseley, A.E.; Deepak, M.S.; Mathers, L. **1995**. Use of methyl bromide and the economic impact of its proposed ban on the Florida fresh fruit and vegetable industry. University of Florida, Agricultural Experiment Station, Institute of Food and Agricultural Sciences Technical Bulletin 898. 204 pp.
61. Barse, J.R.; Ferguson, W.; Seem, R. **1988**. Economic effects of banning soil fumigants. United States Department of Agriculture Economic Research Service. Agricultural Economic Report number 602.
62. Ferguson, W.; Padula, A. **1994**. Economic effects of banning methyl bromide for soil fumigation. United States Department of Agriculture Economic Research Service, Agricultural Economics Report number 677.
63. Dunn, R.A.;Noling, J.W. **1995** Florida Nematode Control Guide. University of Florida, Institute of Food and Agricultural Sciences, Florida Cooperative Extension Service. SP-54. 167 pp.
64. Rhoades, H.; Beeman, J.F. Proc. Fla. State Hort. Soc. **1967**, 80, 156-161.
66. Radewald, J.D.; McKenry, M.V.; Roberts; P.A.; Westerdahl, B.B. Calif. Agric. **1987**, 41, 16-17.
67. Noling, J.W. J. Nema. **1993**, 24, 114.
67. Pohronezny, K.; McSorley, R. Nematol. Medit. **1981**, 9, 151-157.
68. Schuster, D.J.; Montgomery, R.T; Gibbs, D.L.; Marlowe,Jr., G.A.; Jones, J.P.; Overman, A.J. Proc. Fla. State Hort. Soc. **1980**, 93, 235-239.
69. Geraldson, C.M.; Overman, A.J.; Jones, J.P. Proc. Soil Crop Sci. Soc. Fla. **1965**, 25, 18-24.

Chapter 3

1,3-Dichloropropene Regulatory Issues

D. M. Roby[1] and M. W. Melichar[2]

[1]U.S. Regulatory, Toxicology, and Environmental Affairs, and
[2]U.S. Crops Research and Development,
DowElanco, 9330 Zionsville Road, Indianapolis, IN 46268-1054

Over the last several years, use of 1,3-dichloropropene (1,3-D) has come under increased regulatory scrutiny through initiation of the reregistration and special review processes at the U.S. Environmental Protection Agency as well as justification of continued issuance of use permits within the state of California. Resolution of regulatory issues has required the use of a comprehensive strategy that includes definition of benefits and potential risk associated with 1,3-D use, development of state-of-the-art risk refinement and management technologies, research to define effectiveness of potential elements of risk mitigation, and exposure reduction measures through labeling and enhanced product stewardship efforts.

Telone soil fumigants are currently registered as a pre-plant soil treatment used to protect more than 120 vegetable crops, field crops, and nursery crops as well as planting sites for citrus trees, deciduous fruit trees, nut trees, and berry bushes and vines (1). Telone soil fumigants may be applied as a pre-plant soil treatment to control all economically significant nematodes including the following types of plant parasitic nematodes: burrowing, citrus, cyst (sugar beet, soybean, carrot, and wheat), dagger, lance, reniform, ring, root knot, root lesion (meadow), spiral, sting, and stubby root. Telone soil fumigants can also be used to control garden centipedes (symphylans) and wireworms. Telone soil fumigants can suppress sugar beet Rhizomania disease, Granville wilt of tobacco, Fusarium wilt of cotton, Verticillium wilt of mint and potatoes, and aid in the control of bacterial canker of peaches. Soil fumigation with 1,3-D is a component in integrated pest management programs. These programs can include combinations with both chemical and nonchemical elements. Chemical combinations can include other fumigants such as chloropicrin and metam sodium products and nonfumigants such as contact nematicides. Nonchemical components include crop rotation and resistant varieties.

0097-6156/96/0652-0025$15.00/0
© 1996 American Chemical Society

Special Review and Reregistration

1,3-D was placed in the reregistration and special review processes by the U.S. Environmental Protection Agency (EPA) in 1986 (2). All studies required by various data call-ins have been submitted to EPA. The special review process requires EPA to conduct a risk and benefit analysis and publish a proposed regulatory decision (PD2/3) in the Federal Register for public comment and to obtain input from USDA and the Science Advisory Panel. After consideration of all comments and dependent upon submission of additional data or mitigation measures proposed by the registrant, EPA will publish its final determination (PD4).

In conducting its risk and benefit analysis, EPA has three options: 1) propose to retain all uses currently on the product label; 2) propose to cancel all uses because the risks outweigh the benefits; or 3) modify the registration to delete uses, add protective measures or modify use patterns. DowElanco believes that significant advances have already been made in risk mitigation and should be sufficient for EPA to allow continued use of 1,3-D (1, 3).

While separate assessments are conducted for residents and workers, factors affecting both residents and worker exposure include crop rotation practices, application rates, number of times 1,3-D is applied per year, type of application (row or broadcast), and soil type and condition (temperature, moisture, organic content, tilth). Factors affecting worker exposure include engineering controls such as personal protective equipment and the tasks being performed. Factors affecting resident exposure include proximity to fields, activity patterns, atmospheric conditions such as temperature inversions, and product application and sealing practices.

Several factors will make the 1,3-D PD2/3 different from other EPA assessments. They include the fact that residential inhalation risk is a new issue for EPA's Office of Pesticide Programs (OPP). EPA is conducting a geographical analysis looking at localized risk and benefit differentials. The registration status of alternatives are being considered and it is recognized that 1,3-D is one of a few remaining fumigants. This leads to significant benefits of use which are documented in some 46 volumes containing nearly 10,000 pages of use and benefits information submitted to EPA in response to the requirements of data call-ins (4, 5).

Residues

Studies with radiolabeled 1,3-D have shown that the compound is rapidly degraded and the radiolabeled carbon incorporated into natural plant products (6, 7). No residues of concern have been identified in the crops grown in soil fumigated with 1,3-D.

A metabolism study with radiolabeled 1,3-D in lactating goats demonstrated that 1,3-D and/or its metabolic products were rapidly excreted or expired following multiple dosing at 1300 times the potential dietary exposure level for five days (8).

A poultry metabolism study with radiolabeled 1,3-D demonstrated that 1,3-D and/or its metabolic products were rapidly excreted or expired following multiple doses at 3500 times the potential dietary exposure level for seven days (9). The terminal residue in laying hens presented no toxicological concern.

Field residue trials (*10-14*) have confirmed the absence of residues of 1,3-D in carrots, onions, grapes, cantaloupe, broccoli, lettuce, potatoes, pineapples, sugar beets, soybeans, oranges, peaches, cottonseed, and peanuts at a limit of quantitation of 0.01 μg/g. Therefore, no raw agricultural commodity tolerances are required.

Due to the fact that there are no detectable 1,3-D residues in crops following soil fumigation with 1,3-D, there is no dietary exposure issue. Therefore, the only relevant route of exposure for consideration of human risk incidental to agricultural use of 1,3-D is inhalation.

Environmental Fate

Aerobic soil half-lives of 1,3-D vary with respect to soil types, ranging from 1.7 to 53 days (*15*). 1,3-D is converted to naturally occurring carboxylic acids and to CO_2. 1,3-D and its metabolites become increasingly associated with the soil matrix with time. 1,3-D has a vapor pressure of 28 mm/Hg, water solubility of 2 g/L and a hydrolysis rate of 3 days (30 °C) to 51 days (10 °C) depending upon temperature conditions (*16*).

The maximum depth of detectable residues in two field dissipation studies was less than 10 feet. This movement was due to diffusion rather than leaching and half- life values range from 0.6 to 84 days (*17, 18*).

1,3-D reacts with sunlight in the presence of free hydroxyl radicals with an estimated air photolysis half-life of 7 to 12 hours.

In field volatility studies (*16*), 1,3-D flux was shown to vary as a function of soil moisture and temperature conditions, depth of injection, quality of soil sealing, application rates, time-of-day and number of days after application. Peak emissions occurred during late afternoon and early evening periods and a mass loss ranging from 11 to 26% of applied 1,3-D occurred 2 to 5 days following application.

Ground Water

The potential for 1,3-D to contaminate ground water is very low due to several modes of dissipation including, gaseous diffusion throughout the soil, flux through the soil/air interface, hydrolysis in water, and biological metabolism by aerobic and anaerobic microorganisms. The historical incidence of contamination is very low. There were only 6 detections out of 21,270 wells sampled as stated in EPA's Ground Water Data Base of 1992 (*19*). All detections were less than the Maximum Contamination Level of 0.5 ppb.

Toxicology

DowElanco believes that the data base for 1,3-D demonstrates the lack of genotoxic activity under normal physiological conditions *in vivo* (*20-32*). There has been no evidence of teratogenicity (*33*) or effect upon reproduction (*34, 35*), even at toxic exposure levels.

In older National Toxicology Program chronic studies (*36*) (oral bolus) malignant forestomach tumors and an increased incidence of benign liver tumors were reported in rats, and in mice, malignant tumors in urinary bladders and forestomachs and benign tumors in lungs were observed. An inhalation bioassay of 1,3-D (37) conducted in rats and mice reported an increased incidence of benign lung tumors in male mice only. No other tumors related to the inhalation of 1,3-D were reported.

In new chronic dietary feeding studies (38, 39) (by encapsulation), there was no indication of carcinogenic response in mice. A statistically-identified increase in benign liver tumors was identified in high dose group male and female rats. Numbers of benign liver tumors were also increased relative to the historical control incidences of this tumor type in males at the middle dose level. No other tumorgenic response was observed.

A series of studies (40-44) have been conducted to determine the pharmacokinetic behavior and metabolism of 1,3-D in rats, mice and humans. Data indicated that 1,3-D was absorbed from the skin, respiratory tract and gastrointestinal tract. Following absorption, both cis- and trans-isomers of 1,3-D were rapidly eliminated from the bloodstream of rats (half-life approximately 2-4 minutes) and humans (half-life < 10 minutes).

Product Safety

Risk Mitigation. Several advances in occupational and off-site risk mitigation have been implemented (1, 3) including: personal protective equipment, management of end row spillage, product application rate reductions, soil sealing improvements, soil moisture management, dry disconnects and vapor recovery during bulk product transfers and planned phase out of drums by December 31, 1996.

Human Risk Assessment and Refinement. Recent advancements in 1,3-D human exposure assessment including probabilistic (Monte Carlo) methods of exposure assessment and air dispersion (ISCST) modeling have allowed refinements in estimates of exposures encountered by occupational populations and populations that reside in areas of agricultural 1,3-D use (16, 45, 46). Both measured and estimated 1,3-D air concentrations for occupational and residential populations compared to the U.S. EPA IRIS Reference Concentration (20 $\mu g/m^3$) indicate that these populations are likely to be without appreciable risk of health effects during a lifetime. Further, the calculated, hypothetical cancer risks for these populations indicate that lifetime cancer risks are in the range of 1 X 10^{-5} to 1 X 10^{-8} which are generally considered to be acceptable.

The California Experience

On April 13, 1990, the California Department of Pesticide Regulation ceased issuance of use permits for Telone soil fumigants as a result of the Air Resources Board (ARB) monitoring in Merced county (47). DowElanco was allowed to conduct further studies including air monitoring and modeling, to help determine how the meassured air concentrations related to relevant pesticide exposure and risk considerations before reintroduction of Telone was allowed to occur. Since this was the first time action had been taken because of air pathway concerns and pesticide exposure, we had to maintain excellent communications as well as utilize strong technical skills.

A strategy was developed that encompassed significant technology advances, frequent and consistent communications, identification of "real world" solutions that minimized 1,3-D loss to the atmosphere while being economically feasible for producers, and public affairs elements to manage expectations.

Technologies needed to be developed to more accurately assess potential resident exposures from agricultural sources, validate them, and then use them to measure the effectiveness of exposure mitigation efforts implemented in the field. The new technology was developed over a three year period and included risk refinement techniques such as computer modeling using many years of product usage and weather data, and Monte Carlo analysis incorporating use of such variables as breathing rate and residence time.

Every aspect of product management was examined, including conditions of application, crop and acre usage guidelines, utilization of a single distributor and custom application. Application parameters such as soil moisture, soil temperature, soil preparation, tilth, and depth of application were considered. Many of the risk management strategies developed are currently possible only in California. As example, no other state has a tracking system that documents land use on a county, township, section and individual field basis.

After three years of development, the new technology needed to be tested to determine whether what it predicted could be validated by actual measured air concentrations under commercial use conditions. The Monterey Project (*48*) conducted in the fall of 1993 was designed to validate the technological assumptions through simulation of commercial applications under revised management systems. The results showed excellent agreement between predicted and measured air concentrations. Independent air sampling by ARB agreed with DowElanco results and proved that the new technology worked on a commercial scale.

The next step was to assure all outstanding data, risk assessments, an analysis of the commercial validation project, and a proposal for commercial reentry was submitted to the California Department of Pesticide Regulation (DPR) in December 1993. This phase of the process was necessary to define the scope of future commercial opportunities so that California authorities could appropriately determine the acceptability of potential risk using the tools that had been developed over three years and validated in the Monterey Project.

After a year of regulatory review, refinement of risk assessments and agreement of permit conditions, the California DPR recommended reinstatement of Telone II use permits in 13 counties on December 7, 1994 (*49*). The first application of 1,3-D since April 13, 1990 in California occurred on March 3, 1995 in Merced county. A phased reintroduction is underway and proposals for future use expansion are under consideration. While the California experience was long, involved and intense, it significantly increased our effectiveness and demonstrated out ability to work through complex issues with regulators. The success of this reintroduction was possible only through an effective partnership with California policy makers and scientists. The new technology developed for air monitoring, risk assessment and modeling are revolutionary for the industry and will lead the way to the future.

References

1.   DowElanco Product Label for Telone II, **1995.**
2.   Moore, J.A., EPA Letter to Dow Chemical U.S.A. Initiating the Special Review for 1,3-D. Federal Register Notice, September 30, **1986.**
3.   Smith, L.L., DowElanco Letter to EPA, Request to Amend Telone Registrations, October 7, **1992.**
4.   Dow Chemcial Response to EPA Data Call-In. **1987.**
5.   DowElanco Response to EPA Data Call-In. **1991.**
6.   Barnekow, D.E. **1993.** Unpublished results of DowElanco.
7.   Barnekow, D.E. **1993.** Unpublished results of DowElanco.
8.   Satonin, D.K., **1994.** Unpublished results of DowElanco.
9.   Hamburg, A.W., **1993.** Unpublished results of DowElanco.
10. Dixon-White, H.E., **1994.** Unpublished results of DowElanco.
11. Dixon-White, H.E. and D.K. Ervick, **1991.** Unpublished results of DowElanco.
12. Dixon-White, H.E. and T.A. Fasbender, **1991.** Unpublished results of DowElanco.
13. Glas, R.D. **1980.** Unpublished results of The Dow Chemical Company.
14. Glas, R.D. **1981.** Unpublished results of The Dow Chemical Company.
15. Batzer, F.R., J.L. Balcer, and J.D. Wolt. 210th American Chemical Society National Meeting Proceedings, August, **1995.**
16. Knuteson, J.A. and D.G. Petty. 210th American Chemical Society National Meeting Proceedings, August, **1995.**
17. Knuteson, J.A. **1992.** Unpublished results of DowElanco.
18. Knuteson, J.A. **1993.** Unpublished results of DowElanco.
19. U.S. Environmental Protection Agency, *Pesticides In Ground Water Database: A Compilation Of Monitoring Studies: 1971-1991, National Summary,* **1992.**
20. Climie, I. Hutson, D., Morrison, B., Stoydin, G. *Xenobiotica* **1979,** 9, 149-156.
21. Creedy, C., Brooks, T., Dean, B., Hutson, D., Wright, A., *Chem.-Biol. Interactions* **1984,** 50, 39-48.
22. De Lorenzo, F., Degl'Innocenti, S., Ruocco, A., Silengo, L., Cotese, R., *Cancer Res.* **1977,** 37, 1915-1917.
23. Eder, E., Neudecker, T., Lutz, D., Henschler, D., *Chem. Biol. Interact.* **1982,** 38, 303-315.
24. Gollapudi, B.B., Bruce, R.J., Hinze, C.A., **1985.** Unpublished results of The Dow Chemical Company.
25. Mendrala, A.L., **1985.** Unpublished results of The Dow Chemical Company.
26. Mendrala, A.L., **1986.** Unpublished results of The Dow Chemical Company.
27. Neudecker, T., Stefani, A., Henschler, D., *Experientia* **1977,** 33, 1084-1085.
28. Schiffman, D., Eder, E., Neudecker, T., Henschler, D., *Can. Lett.* **1983,** 20, 263-269.
29. Shelby, M.D., Erexson, G.L., Hook, G.J., Tice, R.R., *Environ. Mol. Mutag.* **1993,** 21, 160-179.
30. Stolzenberg, S., Hine, C., *Environ. Mut.* **1980,** 2, 59-66.
31. Talcott, R., King, J., *J. Natl. Cancer Inst.* **1984,** 72, 1113-1116.

32.   Watson, W.P., Brooks, T.M., Huckle, K.R., Hutson, D.H., Land, K.L., Smith, R.J., Wright, A.S., *Chem. Biol. Interact.* **1987**, 61, 17-30.

33.   Hanley, T.R., John-Greene, Young, J.T., Calhoun, L.L., Rao, K.S., *Fundam. Appl. Toxicol.* **1987**, 8, 562-570.

34.   Breslin, W.J., Kirk, H.D., Streeter, C.M., Quast, J.F., Szabo, J.R., **1987.** Unpublished results of The Dow Chemical Company.

35.   Linnett, S.L., Clark, D.G., Blair, D., Cassiday, S.L., *Fundam. Appl. Toxicol.* **1988**, 10, 214-223.

36.   National Toxicology Program, NTP Tech. Report No. 269 **1985**, Government Printing Office, Washington, D.C.

37.   Lomax, L.G., Stott, W.T., Johnson, K.A., Calhoun, L.L., Yano, B.L., Qaust, J.F., *Fundam. Appl. Toxicol.* **1988**, In Press.

38.   Redmond, J.M., Stebbins, K.E., Stott, W.T., **1995**, Unpublished results of The Dow Chemical Company.

39.   Stott, W.T., Johnson, K.A., Jeffries, T.K., Haut, K.T., Shabrang, S.N., **1995**, Unpublished results of The Dow Chemical Company.

40.   Dietz, F., Dittenber, D., Kirk, H., Ramsey, J., *The Toxicologist* **1984a**, 4, 147, Abstract No. 586.

41.   Dietz, F., Hermann, E. Ramsey, J., *The Toxicologist* **1984b**, 4, 147, Abstract No. 585.

42.   Hutson, D., Moss, Pickering, B., *Fd. Cosmet. Toxicol.* **1971**, 9, 677-680.

43.   Stott, W.T., Kastl, P.L. *Toxicol. Appl. Pharmacol.* **1986**, 85, 332-341.

44.   Waechter, J.M., Brazak, K.A., McCarty, L.P., LaPack, M.A., Brownson, P.J. *The Toxicol.* **1992**, 13, Abstract No. 1090.

45.   Calhoun, L.L. **1994**. Unpublished results of DowElanco.

46.   Houtman, B.A. **1993**. Unpublished results of DowElanco.

47.   Telone (1,3-dichloropropene) Monitoring in Merced County. Test Report No. C90-014. Report Date: January 4, **1991**.

48.   Oshima, R.J. California Department of Pesticide Regulation Letter Dated July 23, **1993,** to DowElanco Approving the *DowElanco Telone II Commercial Use Project in Monterey County.*

49.   California Department of Pesticide Regulation, News Release Dated December 7, **1994**, Release No. 94-42, *DPR Approves Limited Use of Soil Fumigant.*

Chapter 4

Hydrolysis of Methyl Bromide, Ethyl Bromide, Chloropicrin, 1,4-Dichloro-2-butene, and Other Halogenated Hydrocarbons

Peter M. Jeffers[1] and N. Lee Wolfe[2]

[1]Chemistry Department, Bowers Hall, P.O. Box 2000,
State University of New York, Cortland, NY 13045
[2]Ecosystems Research Division, National Exposure Research Laboratory,
U.S. Environmental Protection Agency, 960 College Station Road,
Athens, GA 30605–2700

The hydrolysis of halogenated hydrocarbons in general, and of four soil fumigants (methyl and ethyl bromide, chloropicrin, and 1,4-dichloro-2-butene), in particular, is reviewed. Experimental methods are discussed, predictive generalizations are presented, and new experimental data on the hydrolysis of the fumigants are given. Some semi-quantitative results are mentioned concerning adsorption and enzymatic degradation of methyl bromide and chloropicrin by plant materials. At 25 °C, the hydrolysis half-lives of both methyl and ethyl bromide are about 21 days, while at 35 °C this value is about 5 days. The 25 ° C hydrolysis half-life for dichlorobutene is about 3 hours, but chloropicrin has an extrapolated half-life of 500,000 years. Enzymatic dehalogenation of methyl bromide and chloropicrin provides a transformation pathway that results in environmental half-lives of less than 20 hours.

Early studies of the hydrolysis rates of halogenated hydrocarbons date from at least 60 years ago. Sir C. N. Hinshelwood of Oxford University reported in 1933 that ethyl chloride eliminates HCl only by an SN_2 process (1). E. A. Moelwyn-Hughes and his students at Cambridge published a set of classic chemical kinetics papers on the hydrolysis of halogenated methanes (2-5) between 1941 and 1959, identifying neutral and alkaline reaction pathways, products, and mechanisms. Much of this early work remains valid even though it predated the simplicity, accuracy, and speed of modern chromatographic analytical methods.

The most useful "modern" measurements of halocarbon hydrolyses date from 1974 when R. Walraevens and co-workers reported on the elimination of HCl from chlorinated ethanes under alkaline conditions (6). W. L. Dilling's research group presented hydrolysis/volatilization loss of various chlorinated hydrocarbons with an environmental flavor in 1975 (7), but their data were limited to measurements at room temperature. Since that time, numerous research papers have appeared that are concerned with the hydrolysis of various halogenated compounds, and excellent review papers are available (8-11).

0097–6156/96/0652–0032$15.00/0

"True" environmental half-lives may be overestimated by orders of magnitude if only homogeneous hydrolysis processes are considered. Horvath (*12*) suggested in 1972 that bacteria can co-metabolize halocarbons, Barrio-Lage *et al.* (*13*) demonstrated sequential dehalogenation of the ethenes by anaerobic bacteria in 1985, and Jafvert and Wolfe (*14*) reported rapid and apparently abiotic degradation of haloethanes in sediment-water systems in 1987.

Experimental Methods

T. Mill, *et al.*, (*15*) published an "environmentalist's protocol" for determining the fate of organic chemicals in air and water, suggesting hydrolysis measurements at temperatures of 25, 35 and 55° C and with solutions adjusted to pH 3, 7, and 11. J. J. Ellington and co-workers followed this prescription to evaluate the environmental half-lives of 80 different compounds between 1985 and 1987 (*16,17*).

Assisted by SUNY Cortland undergraduates, we have measured and reported hydrolysis rates for 32 halogenated hydrocarbons (*18,19*), using the more pragmatic approach of finding conditions where reaction rates can be determined readily. General experience indicates that halocarbons do not undergo hydrolysis by an acid-moderated pathway; thus, we designed our experiments to measure the "neutral" hydrolysis rate in solutions 0.01 molar in HCl, and followed the alkaline hydrolysis in dilute solutions of NaOH, rather than using pH buffers. We worked within temperature and pH ranges that yielded experimental half-lives of 10 min. to 50 hours by adjusting pH with the dilute acid or base from 2 to 13, and temperatures from 0 to 190 °C. These conditions are unusual, from an environmentalist's perspective, but they are not unique to our studies. For instance, in 1959 Moelwyn-Hughes followed carbon tetrachloride hydrolysis at temperatures up to 150 °C (*4*). We achieved the high temperatures by sealing aqueous samples in small Pyrex glass bulbs (*20*), a set of which was heated to the desired temperature and sampled over an appropriate time interval.

Our rate measurements were designed to allow determination of both the neutral and alkaline rate constants, k_N and k_B, in the expression

$$\text{Rate} = -d(X)/dt = k_N(X) + k_B(OH^-)(X)$$

where (X) represents the molar concentration of the halocarbon and (OH⁻) is the molar hydroxide concentration. Both rate constants are assumed to fit the Arrhenius form

$$k = Ae^{-(E/RT)}$$

where A is the pre-exponential or frequency factor, and E is the activation energy, R is the gas constant in appropriate energy units, and T is the absolute temperature. We worked with NaOH concentrations that were sufficiently larger than reactant concentrations that pseudo first-order behavior could be assumed. For the data and results listed in this paper, time units are minutes, concentrations are mole/l, and energies are in kJ/mole.

Methyl and Ethyl Bromide Experiments. Solutions were prepared that contained methyl bromide, ethyl bromide, and trichloroethene (TCE), each at a concentration of about 10^{-4} mol/l. The ethyl bromide was dissolved as the pure liquid, then diluted.

Methyl bromide was passed into the solution as the pure gas from a gas-tight syringe. TCE was diluted from a saturated aqueous solution. The reaction solution was pipetted immediately into 2 ml autosampler vials with zero headspace. The vials were capped and stored at 4 °C until use. Previous experiments indicated that monohalogenated compounds do not react by an alkaline pathway at a pH below 10, so no attempt was made to adjust the pH of these solutions. The TCE was included as a non-reactive internal standard. Kinetics runs were performed from 35 °C to 85 °C with time intervals ranging from 135 hours for 35 °C experiments to 94 min. at 85 °C.

Chloropicrin Experiments. Chloropicrin, CCl_3NO_2, has sufficient structural similarity to carbon tetrachloride that we expected it to react only by a neutral mechanism. Solutions of chloropicrin at concentrations of about $3x10^{-4}$ mol/l were prepared by dissolving the pure liquid in deionized water. Hydrolysis experiments were attempted at temperatures from 85 to 166 °C, but measurable extents of reaction were observed only above 140 °C. Analysis was by GC/MS and by ion chromatography.

1,4-Dichloro-2-butene Experiments. Dichlorobutene solutions were prepared by dilution of a saturated solution, and contained TCE as internal standard. These solutions were not adjusted for pH, and hydrolysis rates were determined at 23, 49, and 70 °C.

Qualitative Enzymatic Degradation Experiments. Two 50 ml beakers were filled with a 10^{-4} mol/l solution of methyl bromide. To one beaker was added 2.5 g of the aqueous plant *Myriophyllum aquatica* (parrot feather), and both beakers were covered with Parafilm and sampled by GC/MS for 2 days. A similar experiment utilized samples prepared directly in sealed 2-ml autosampler vials, comprising a "blank" methyl bromide solution, a vial with parrot feather added, and a third vial containing leaves of the *Arctostaphylos uva-ursi* (bear berry) bush, which is known to be high in dehalogenase enzyme (32).
    Another set of autosampler vials was prepared with blank, cut up pieces of rice plant, and parrot feather leaves, with a much heavier "loading" of the parrot feather. Finally, two autosampler vials were packed with trefoil root and crabgrass root, then filled with the methyl bromide solution, and sampled with time. All these vials were sealed immediately after filling, and were sampled by puncturing the teflon-faced septum.
    A set of 2.5-ml screw cap vials containing 0.5 g parrot feather each were filled with a $6x10^{-4}$ molar solution of chloropicrin. These vials were sampled over 20 hours and were analyzed both by GC/MS and by ion chromatography.

Results and Discussion

We begin by citing some reactivity correlations observed in our previous publications (*18,19*).
    * Both neutral and alkaline hydrolysis processes must be considered. For example, while carbon tetrachloride and 1,1,1-trichloroethane hydrolyze only by neutral routes, the major degradation of 1,1,2-trichloroethane, 1,1,2,2-tetrachloroethane, and pentachloroethane is by the alkaline pathway, even at pH 7. In naturally alkaline waters of pH 9-10, elimination of HCl from 1,2-dichloroethane to give vinyl chloride becomes competitive with the neutral mechanism product, ethylene glycol.

* Alkaline reactivity increases with proton acidity, and there is some correlation with experimentally determined activation energy, in that reactions with lower E clearly have faster rates.

* With the neutral reactions there is a complex interplay of steric and energetic factors. Both elimination and substitution are observed, and there is not always a single mechanism operating.   With 1,1,1-trichloroethane, both acetic acid and 1,1-dichloroethene are produced.

* There is an enormous range of reactivity. At a given temperature, the neutral rate constants vary by a factor of 10^7, and the alkaline rate constants cover a range differing by factors as great as 10^{12}. Environmental hydrolysis lifetimes of halogenated hydrocarbons range from days to geologically significant spans.

* Although alkaline hydrolysis of singly halogenated compounds can be observed experimentally, the neutral reaction dominates in solutions any less concentrated than pH 10.

* Not all factors dictating reactivity are obvious. There is still room for experimental work.

Some additional predictive generalizations became obvious from our studies with change of halogen substituent.

** A brominated compound is more reactive than its chlorinated analog by a factor of 10-100, by both neutral and alkaline reaction mechanisms.

** Greater reactivity of brominated versus chlorinated compounds is due to both lower activation energies and entropy effects (as reflected in the preexponential factor) that generally contribute in the same direction.

** Within a series of halogenated compounds, the order of reactivity is identical for brominated and chlorinated analogs.

** Substitution of fluorine for chlorine in the ethanes causes an enormous drop in neutral hydrolysis reactivity due to solvation rather than bond energy effects.

** Substitution of fluorine for chlorine in the ethanes greatly retards alkaline hydrolysis due to large increases in activation energy.

Methyl and Ethyl Bromide Homogeneous Hydrolysis. Figures 1 and 2 show our results for the homogeneous neutral hydrolysis rate constant determinations for methyl and ethyl bromide. The data for Figure 1 are a combination of values extracted from Moelwyn-Hughes paper (*4*) and our own experiments. Regression of the entire data set for methyl bromide yields

$$k_N(CH_3Br) = (3.4 \pm 0.5) \times 10^{13} \, e^{-(103,500 \pm 2,000)/RT} \, min^{-1}, \quad r^2 = 0.997.$$

Regression analysis of the data in Figure 2 for ethyl bromide shows

$$k_N(C_2H_5Br) = (5.7 \pm 1.1) \times 10^{13} \, e^{-(105,000 \pm 3,200)/RT} \, min^{-1}, \quad r^2 = 0.992.$$

From these rate constants, we calculate 25 °C hydrolysis half-lives of 20.1 and 21.9 days for methyl and ethyl bromide, respectively, while at 35 °C the half-lives are calculated to be 5.2 and 5.6 days. These results are very close to the original results reported by Moelwyn-Hughes, as cited by Mabey and Mill (*8*). It is interesting to note that the rate

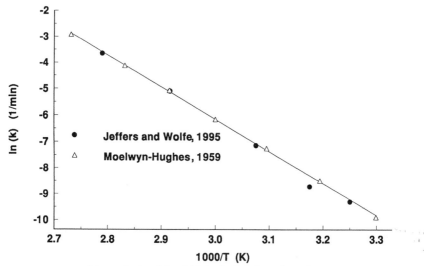

Figure 1. Combined historical and modern data.

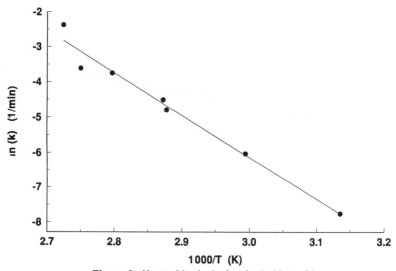

Figure 2. Neutral hydrolysis of ethyl bromide.

parameters for methyl and ethyl bromide are nearly within experimental uncertainty of one another.

Dichloro-2-butene Homogeneous Hydrolysis.   Experiments on the neutral, homogeneous hydrolysis of 1,4-dichloro-2-butene at 23 , 49, and 70 °C indicated a rate constant of:

$$k_n(DCB) = (3.0\pm1.1) \times 10^9\ e^{-(67,900\pm7,900)/RT}\ min^{-1}.$$

From this rate constant expression, a 25 °C half-life of 3 hours is calculated.

Chloropicrin Homogeneous Hydrolysis.   Chloropicrin is extremely resistant to homogeneous hydrolysis, although it is a very reactive and toxic compound. Our initial attempts to observe hydrolysis at 85 °C were rewarded with no apparent reaction after 40 hrs. Attempts to measure hydrolysis at 124 °C using TCE as a trace internal standard were confounding, since the TCE, which should be entirely inert under these conditions, disappeared, while the chloropicrin remained. Figure 3 shows our measured hydrolysis rate constants for chloropicrin from 124 to 166 °C, with analysis done in some cases by GC/MS and in other instances by ion chromatography. With the ion chromatographic analysis, we observed that chloride ion (corrected for the 3:1 stoichiometry) appeared at least twice as fast as nitrate ion, implying that the total degradation is a complex and multi-step process. We estimate a minimum activation energy of 162 kJ/mol, and an A-factor of $1.9\times10^{17}\ min^{-1}$. That these activation energy and frequency factor values are so much higher than any others we have ever measured may mean the hydrolysis process for chloropicrin is mechanistically unlike that for "normal" halocarbons. The extrapolated half-life at 25 °C for the activation energy mentioned above implies that homogeneous hydrolysis is completely negligible as an environmental process for chloropicrin.

Plant Enzyme Studies.   In 1987, Jafvert and Wolfe *(14)* reported rapid degradation of halogenated ethanes in anaerobic sediments, and Weber and Wolfe *(21)* found reduction of aromatic azo compounds in sediments.   These sediment-moderated reductions appeared to be abiotic processes. Wolfe and Jeffers *(22)* and Peijnenburg *et al.* were able to construct Structure Activity Relationships (SAR's) for aliphatic halocarbons *(23)* and for halogenated aromatic halocarbons *(24)*. The success of these SAR's and the fact that sediment activity scaled linearly with sediment organic content was strong circumstantial evidence supporting the abiotic nature of the reduction reactions.   In 1992, Wolfe and Macalady *(25)* reviewed the chemistry of abiotic transformations of organic pollutants in anaerobic ecosystems, finding that, although the natural reductants remained elusive, evidence was strong for widespread reductive mediation of reactions by bio-organic molecules.   Along these lines, Gantzer and Wackett *(26)* reported reductive dechlorination to be catalyzed by bacterial transition metal co-enzymatic porphyrin compounds. By 1993, Masunaga, Wolfe and Carriera *(27)* were able to extract a protein from sediment that caused benzonitrile transformations similar to that of nitrilase. These sediment enzymes are sufficiently robust to have

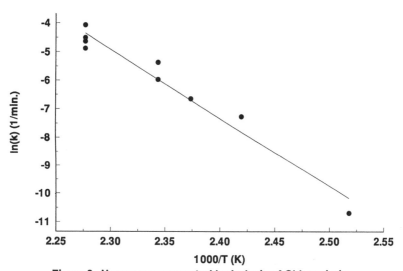

Figure 3. Homogeneous neutral hydrolysis of Chloropicrin.

relatively long, active chemical lifetimes in the anoxic medium. By 1995, our research group at the EPA lab was able to report the isolation of five enzymes from sediments, a dehalogenase, a nitroreductase, a peroxidase, a laccase, and a nitrilase, all presumably originating from plants that had decomposed (*28*). This last paper was a major didactic presentation of phytoremediation, the treatment of various kinds of pollutants by live plants. A paper by Nzengung, Wolfe, Carreira, and McCutcheon (*29*) carries these findings to the logical end of demonstrating that halogenated ethenes are readily reduced by aquatic plants themselves, with no sediment or attendant bacteria.

Our several sets of experiments with methyl bromide solutions in intimate contact with various plant leaves and stems are predicated on the existence of a dehalogenase within green plants that is available to react with halocarbons in surrounding solutions. These rather crude experiments were all consistent with a degradation half-life of less than 20 hours. When we placed methyl bromide solution in vials heavily loaded with root material and soil, there was immediate "adsorption" of at least 90% of the methyl bromide (determined from the GC/MS response with samples of the reactant solution and of solution-plant vials measured immediately after vial proparation), and it could not be extracted with acetonitrile.

Chloropicrin at a concentration of 6×10^{-4} mole/l, in contact with 0.5 g parrot feather plant in a 2.5 ml screw-cap vial, degraded with a half-life under 20 hours. The degradation was a reduction process, evidenced by the observation (GC/MS) of $CHCl_2NO_2$ as the reaction proceeded. The $CHCl_2NO_2$ was an intermediate, since its concentration rose, then fell, never exceeding the parent compound concentration. Ion chromatographic observation of the phytoreduction was confounded by the release of an entire spectrum of ions by the parrot feather, in addition to those from degradation of the chloropicrin. Final Cl^- concentrations were as much as 5 to 10 times as great as the original chloropicrin concentration. In contrast, with homogeneous hydrolysis, there was an excellent mass balance of chloride formed and chloropicrin degraded.

Structure Activity Relations. The quantitative SAR's mentioned above (*22-24*) utilize molecular descriptor parameters taken from Hansch and Leo (*30*) and are constructed along the lines suggested by Shorter (*31*). Wolfe and Jeffers (*22*) include results for 1,2-dibromoethane which has an estimated half-life of 21.2 hr in sediment. Using the same set of molecular parameters, we calculate a sediment half-life of 55 hr for methyl bromide. We extend this comparison by mentioning that the SAR estimates for sediment half-lives of perchloroethylene and carbon tetrachloride are 285 and 105 hr, respectively, and that Nzengung *et al.* (*29*) observed half-lives of 120 and 27 hr, respectively, for these compounds in contact with aqueous plants. Thus we expected that methyl bromide would have a half-life in solutions in contact with green plants rich in dehalogenase that was shorter than, (perhaps considerably shorter than) the 55 hour SAR estimate. As one further extension of these lines of thought, we might expect methyl bromide to be degraded actively by green plants in the gas phase. This kind of reactive sink might be an important inclusion in methyl bromide atmospheric modeling. Further experiments along these lines are reported in a companion paper in this volume.

General Observations. Our experiments are consistent with historic reports that the room temperature hydrolysis half-life of methyl bromide is on the order of 20 days. However, in sun-baked fields, soil temperatures are likely to exceed 25 °C significantly, and the hydrolysis half-life will consequently be considerably shorter. In addition, the enzymatic dehalogenation of methyl bromide by plant materials plowed into fields to be fumigated should lead to much shorter actual lifetimes of the fumigant, and strong adsorption to root materials and soil can provide a long enough soil residence at high enough temperatures to mitigate the amount of methyl bromide escaping to the atmosphere. Details of the fumigation process, including depth of injection and plastic film mulching, are certainly important parameters in limiting the initial escape of the fumigant to the atmosphere. Further careful, quantitative experiments of the various fate pathways of methyl bromide used as a fumigant are warranted.

The compound 1,4-dichloro-2-butene is a close analog to 1,3-dichloro-2-propene. Our hydrolysis experiments indicate that DCP should degrade by hydrolysis within several hours at normal temperatures. Thus use of DCP should pose a minimal environmental threat.

The hydrolysis of chloropicrin is insignificant environmentally. However, plant dehalogenases will degrade chloropicrin readily and completely, within 20 hours, at "reasonable" concentrations. The final products are chloride ion, nitrate ion, and probably carbon dioxide.

Acknowledgments. Parts of this work were completed while PMJ was a National Research Council Senior Fellow at the USEPA, Athens, Ga., where this specific research was funded in part by the Strategic Environmental Research and Development Program (SERDP). The following SUNY Cortland students contributed to this research: Lisa Ward, Lisa Woytowich, Scott Luczak, Paul Coty, Christine Brenner, Kellie Wilson, John Snyder, Bill Smith, Jason Wright, David Haley, Vincent Venezia, Vanessa Rich, Brian Jones, Rod Jones, and Ed Mereand. The research at Cortland was funded, in part, by a grant from the U.S. Department of Agriculture.

Literature Cited

1. Grant, G. H.; Hinshelwood, C. N. *J. Chem. Soc.* **1933**, 258.
2. Moelwyn-Hughes, E. A.; *Proc. Royal Soc. London* **1949**, *A196*, 540.
3. Moelwyn-Hughes, E. A.; *Proc. Royal Soc. London* **1953**, *A220*, 386.
4. Fells, I.; Moelwyn-Hughes, E. A.; *J. Chem. Soc.* **1958**, 1326.
5. Fells, I.; Moelwyn-Hughes, E. A.; *J. Chem. Soc.* **1959**, 398.
6. Walraevens, R.; Trouillet, P.; Devos, A.; *Int'l. J. Chem. Kinetics* **1974**, 6, 777.
7. Dilling, W. L.; Tefertiller, N.; Kallos.; *Environ. Sci. Technol.* **1975**, *9*, 833.
8. Mabey, W.; Mill, T.; *J. Phys. Chem. Ref. Data* **1978**, 7, 383.
9. Vogel, T. M.; Criddle, C. S.; McCarty, P. L.; *Env. Sci. Technol.* **1987**, *21*, 722.
10. Roberts, A. L.; Jeffers, P. M.; Wolfe, N. L.; Gschwend, P. M. *Crit. Rev's. Env. Sci. Technol.* **1993**, *23*, 1.
11. Barbee, G. *GWMR* **1994**, *Winter*, 135.

12. Horvath, R.S., *Bact. Rev.* **1972**, *36*, 146.
13. Barrio-Lage, G.; Parsons, F.Z.; Nassar, R.S.; Lorenzo, P.A. *Environ. Sci. Technol.* **1985**, *20*, 96.
14. Jafvert, C.T.; Wolfe, N.L. *Environ. Toxicol. Chem.* **1987**, *4*, 827.
15. Mill, T.; Mabey, W. R.; Bomberger, D. C.; Chou, T. W.; Hendry, D. G.; Smith, J. H. **1982**, EPA/600/3-82/022.
16. Ellington, J. J.; Stancil, F. E.; Payne, W. D. **1986**, Vol. 1, EPA/600/3-86/043.
17. Ellington, J. J.; Stancil, F. E.; Payne, W. D.; Trusty, C. **1987**, Vol. 2, EPA/600/3-87/019.
18. Jeffers, P. M.; Ward, L.; Woytowich, L.; Wolfe, N. L. *Env. Sci. Technol.* **1989**, *23*, 965.
19. Jeffers, P.M.; Wolfe, N.L. *Environ. Toxicol. Chem.* **1996**, *15*.
20. Jeffers, P. M. *J. Chem. Ed.* **1990**, *67*, 522.
21. Weber, E.J.; Wolfe, N.L. *Environ. Toxicol. Chem.* **1987**, *6*, 911.
22. Wolfe, N. L.; Jeffers, P. M. **1990**, EPA/600/M-89/032.
23. Peijnenburg, W.J.G.M.; 't Hart, M.J.; den Hollander, H.A.; van de Meent, D.; Verboom, H.H.; Wolfe, N.L. **1989**, RIVM Report No. 718817002. National Institute of Public Health and Environmental Protection, Bilthoven, The Netherlands.
24. Peijnenburg, W.J.G.M.; 't Hart, M.J.; den Hollander, H.A.; van de Meent, D.; Verboom, H.H.; Wolfe, N.L. *Environ. Toxicol. Chem.* **1992**, *11*, 301.
25. Wolfe, N.L.; Macalady, D.L. *J. Contaminant Hydrology* **1992**, *9*, 17.
26. Gantzer, C.J.; Wackett, L.P. *Environ. Sci. Technol.* **1991**, *25*, 715.
27. Masunaga, S.; Wolfe, N.L.; Carriera, L.H. *Water Sci. Technol.* **1993**, *28*, 123.
28. Schnoor, J.L.; Licht, L.A.; McCutcheon, S.C.; Wolfe, N.L.; Carreira, L.H. *Environ. Sci. Technol.* **1995**, *29*, 318A.
29. Nzengung, V.A.; Wolfe, N.L.; Carriera, L.H.; McCutcheon, S.C. *Environ. Sci. Technol.* (Submitted, 1996).
30. Hansch, C.; Leo, A.J. **1979**, *Substituent Constants for Correlation Analysis in Chemistry and Biology*. John Wiley & Sons, NY.
31. Shorter, J. **1973**, *Correlation Analysis in Organic Chemistry. An Introduction to Linear Free-Energy Relationships*. Clarendon Press, Oxford, UK.
32. L. Carriera, private communication.

Chapter 5

Fate of Methyl Bromide in Fumigated Soils

Todd A. Anderson[1], Patricia J. Rice[2], James H. Cink[2],
and Joel R. Coats[2]

[1]Institute of Wildlife and Environmental Toxicology, Department
of Environmental Toxicology, Clemson University, One Tiwet Drive,
Pendleton, SC 29670
[2]Pesticide Toxicology Laboratory, Department of Entomology,
Iowa State University, Ames, IA 50011–3140

Recent controversy over the potential role of methyl bromide (MeBr) in damaging the ozone layer has spurred interest in increasing our understanding of the transformation and movement of this fumigant after it is applied to soil. Our research indicates MeBr is rapidly volatilized from fumigated soil (within the first 24 hours) and volatility significantly increases with temperature (35° C > 25° C = 15° C) and moisture (0.03 bar > 0.3 bar > 1 bar > 3 bar). Degradation of MeBr, measured by production of bromide ion (Br⁻), was also directly related to temperature and moisture. Undisturbed soil column studies indicated that MeBr rapidly volatilized (> 50% of the MeBr flux occurred in 48 h) but did not leach into subsurface soil. Residual MeBr was degraded in the soil column, evident by the high concentrations of Br⁻ in the leachate water. In field studies, MeBr also volatilized rapidly from soil, but a significant portion of the MeBr was degraded (30% after 2 d). These studies provide pertinent information for assessing the fate of MeBr in soil, which should lead to more informed decisions regulating its use.

Methyl bromide (MeBr) is a biocidal fumigant used to control a broad spectrum of pests and diseases including nematodes, insects, weed seeds, viruses, and fungi. The average annual amount of MeBr used has increased by 7% since 1984; it is currently the fifth most widely used pesticide in U.S. agriculture, and by volume, it is the second most widely applied insecticide in the world (1). Over 55 million pounds of MeBr were used in the U.S. in 1990; approximately 80% was applied as a soil fumigant, and an additional 15% was employed as a fumigant for agricultural commodities (food and packaging materials) and structures.

While the quantity of MeBr released into the atmosphere from natural sources is estimated to be 75%, anthropogenic emissions (primarily through fumigation, chemical manufacturing, and car exhausts) also contribute significantly to the global MeBr budget (2,3). Large quantities of field-applied MeBr (> 80%) have been shown to volatilize into the atmosphere (4). Although man-made emissions represent only a fraction of the total atmospheric MeBr, they disrupt the natural balance of the atmosphere and are believed to contribute to diminishing the ozone layer (5).

Photolysis of MeBr at high elevations (stratosphere) produces bromine radicals. MeBr's atmospheric life-span (18 months) is relatively short compared with chlorofluorocarbons (50 to 100 years), which are being phased out; however, bromine can scavenge ozone 100 times more efficiently than Cl (5). Recent dispute over MeBr's potential to deplete the ozone layer has led the Environmental Protection Agency (EPA) to propose a phasing out of its use. In 1991 the United Nations Environment Program (UNEP) Montreal Protocol committee classified methyl bromide as a Class I ozone depleter. The EPA is responsible for enforcing a phase out of all Class I ozone depleter chemicals by the year 2001 (6,7). Presently, there appear to be few viable alternatives to replace this fumigant, and banning of MeBr may, by one estimate, result in an annual loss of over $1.3 billion to U.S. consumers and producers (6).

Despite extensive use of MeBr, little is known about its fate in soil. Previous research primarily focused on evaluating the toxicity of MeBr and measuring residues on food. MeBr is considered a minor surface and ground water contaminant and a major air contaminant (8). During fumigation, MeBr penetrates into the soil and is partitioned into the liquid, gas, and adsorbed solid phases. Degradation in the soil may occur by abiotic or biotic reactions and include primarily hydrolysis as well as reduction and oxidation reactions (9-12). Degradation products of MeBr include bromide ion, methanol, formaldehyde, hydrobromic acid, and carbon dioxide (13,14).

The persistence, volatility, and degradation of MeBr (and other organic chemicals) in the soil is influenced by chemical properties, soil properties, and environmental conditions. Information on the fate of MeBr under various conditions is needed to make educated decisions involving its use and regulation. We report here laboratory studies addressing the influence of environmental and soil variables on the degradation and volatility of MeBr and a field experiment on MeBr flux and degradation in fumigated soil.

Experimental Procedures

Chemical. Methyl bromide was obtained from Great Lakes Chemical Co. (West Lafayette, IN) and stored as a liquid at -60° C. Pure MeBr was used for analytical standards and fumigation of laboratory samples.

Soil Collection and Treatment. The pesticide-free soil used in the laboratory studies was obtained from the Iowa State University Agronomy and Agricultural Engineering Farm near Ames (Story County), Iowa. Samples were collected, sieved (2.0 mm), and stored in the dark at 4° C until needed. Ten samples were randomly collected from the field and combined for each replicate. Soil was analyzed by standard methods to determine physicochemical properties. The sandy clay loam soil consisted of 53% sand, 29% silt, 17% clay, 3.1% organic matter, and a slightly acidic pH (6.6). In all the studies described below, liquid MeBr was applied to the soil and allowed to incubate (sealed) for 48 h before initial experimental monitoring, to simulate a typical fumigation period.

Volatility Study. Soil (10 g dry weight) was placed in 45-ml glass bottles equipped with polytetrafluoroethylene-lined septa. Moisture was adjusted to 0.03 bar, 0.3 bar, 1 bar, and 3 bar, respectively. MeBr was applied as a liquid at a concentration of 2,733 μg MeBr/g soil (594 g/m^3), which represented a typical structural fumigation rate. This rate was used instead of a typical field fumigation rate because of the difficulty of applying small quantities of this highly volatile compound. Samples were incubated in the dark at 15° C, 25° C, or 35° C. Each treatment

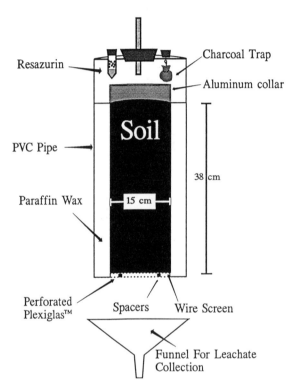

Figure 1.   Undisturbed soil column used to study the fate of methyl bromide in soil.

consisted of four replicates. Concentration of MeBr was measured at intervals using static headspace gas chromatography (15,16) for approximately 5 d (119 h). Headspace above the soil samples was purged with N_2 following each analysis. MeBr flux from soils was determined from headspace concentrations. The data were statistically analyzed using analysis of variance (ANOVA) and least significant difference (LSD) at 5%.

Degradation Study. Fumigated samples were analyzed for Br⁻ to assess the influence of temperature, moisture, and sterility on MeBr degradation. Soil treatments were identical to those previously stated. Sterile control samples (autoclaved >121° C for 30 minutes on three consecutive days) and treatment soils were adjusted to 0.3 bar moisture with sterile deionized water. At periodic intervals, fumigated soils were extracted with 20 mL deionized water by mechanical agitation and centrifugation. The supernatant was removed and analyzed for Br⁻ using a bromide-specific electrode (Orion Research Inc., Boston, MA). Analysis of variance and LSD (5%) were used to determine significant differences between treatments.

Microbial Toxicity. Soil respiration was measured to determine the effect of MeBr on microbial activity. Twenty grams of soil (dry weight) was placed in stoppered, 250-mL glass jars, and soil moisture was adjusted to 0.3 bar. MeBr was applied at concentrations of 594 g/m^3 and 350 $\mu g/g$ to represent structural fumigation and 2.6x the rate of field fumigation, respectively. Soils were incubated in the dark at 25° C. Carbon dioxide efflux was measured at 24-h intervals after the initial 48-h fumigation period. The sample headspace was purged with moist, CO_2-free air and was analyzed using an infrared gas analyzer (Mine Safety Appliances Co., Pittsburgh, PA) (17). Microbial respiration in the fumigated and untreated samples was compared. Treatments were considered significantly different when the SD of the daily means did not overlap.

Soil Column Study. Two undisturbed soil columns (15 cm diameter x 38 cm length) were obtained from an agricultural field site (no previous pesticide history) near Ames, Story County, IA. The procedures for collection and removal of the columns were previously described (18). Additional soil samples were collected at the same depths as the column, and soil physicochemical properties were determined. A composite of the soil samples comparable to the soil column consisted of sandy clay loam soil with a pH of 5.4 and 54% sand, 25% silt, 21% clay, and 2.5% organic matter.

Soil columns (Figure 1) were prepared for laboratory studies as described by Kruger et al. (19). Modifications were made to collect volatilized MeBr from the soil. The PVC pipe surrounding the sides of the column was longer than the soil column to insure sufficient headspace. Four 500-mL increments of deionized water were leached through the columns to determine any background concentrations of Br⁻ and MeBr.

Liquid MeBr was applied to the soil surface and the columns were immediately sealed. Soil columns were incubated for 48 h to allow MeBr to penetrate the soil and mimic fumigation techniques used in the field. The MeBr-fumigated columns were maintained at 24° ± 1° C. Soil columns were leached weekly with 500 mL deionized water to represent 1 inch of rainfall. Leachate was collected at the bottom of the column and analyzed for Br⁻ and MeBr by using a bromide-specific electrode, and gas chromatography, respectively.

Resazurin and granular activated carbon traps were suspended in the headspace of the columns after the 48-h fumigation. Carbon traps consisted of 8 g activated charcoal wrapped in 5 cm x 5 cm, 100% cotton net (1-mm mesh). These traps were changed periodically and used to determine the amount of MeBr in the headspace of the column. Upon removal, the traps were placed in 45-mL glass bottles equipped with screw caps and polytetrafluoroethylene-lined septa and stored at -60° C until analysis. MeBr was desorbed off the carbon traps by the procedure of Woodrow et al. (15) with modifications. Quantities of MeBr detected were considered in the final calculation of MeBr that volatilized from the soil.

At the conclusion of the study, the undisturbed soil columns were cut into 5-cm increments and extracted with water as stated above in the degradation study. The soil extracts were analyzed using a bromide-specific electrode.

Field Study. MeBr fate (volatility, degradation, microbial toxicity) was determined in fumigated field soils. Three adjacent fields near Ames, Story County, IA were professionally treated (by injection) with 390 kg/ha MeBr-chloropicrin mixture (98:2 by weight). The fields were covered with a plastic tarp after injection and remained covered for 48 h. Composite soil samples from the fields were analyzed by standard methods to determine physicochemical properties. The clay loam soil consisted of 43% sand, 29% silt, 27% clay, 3.2% organic matter, and a near neutral pH (7.2).

The volatility of MeBr in the field was determined using glass flux chambers equipped with activated carbon traps. Flux chambers were placed on the plastic tarp immediately following fumigation. Upon removal of the tarp, flux chambers were placed directly on the soil. Carbon traps were collected at various time intervals and stored at -60° C until analysis by headspace gas chromatography.

The concentration of MeBr in the soil gas was also determined. Soil probe samples were collected at various time intervals, equilibrated at room temperature in 100-mL glass bottles, and analyzed by headspace gas chromatography.

Bromide ion concentrations in the soil were measured, following analysis of gaseous MeBr, as an indicator of MeBr degradation. Soil samples (10 g) were extracted with 30 mL distilled deionized H_2O. Extracts were analyzed using an ion-specific electrode.

Microbial respiration of fumigated soils was monitored in order to determine the potential toxicity of 390 kg/ha field-applied MeBr to soil microorganisms. Soil probe samples were randomly collected from the MeBr fumigated fields, and analyzed as described above.

Analysis of Bromide Ion. Supernatant and leachate samples from degradation studies, column studies, and field studies were measured for Br⁻ using a bromide-specific electrode attached to a pH meter (Fisher Scientific, Pittsburgh, PA). Br⁻ standards were prepared with NaBr, deionized water, and 5 M $NaNO_3$ (ionic strength buffer). Calibration curves were constructed from the standards and used to determine the sample concentrations.

Analysis of MeBr. Procedures for the analytical standards and analysis of sample and standard headspace were modified from Woodrow et al. (15). Methyl bromide standards were made in benzyl alcohol, stored at -60° C, and replaced every 2 weeks. Samples were analyzed on a Varian 3700 gas chromatograph equipped with a Ni^{63} electron-capture detector at 350° C. Injector temperature was 170° C with a column temperature of 160° C. The glass column (0.912 m x 2.0 mm i.d.) was packed with 100/120 mesh Porapak Q (Supelco Inc., Bellefonte, PA) on Carbopack with a carrier gas consisting of ultra pure N_2 (26 mL/min).

Results and Discussion

Volatility Studies. Methyl bromide was significantly more volatile in soil samples incubated at 35° C, with no significant difference between the 15° C and 25° C soils (Table 1). The flux of MeBr in 35° C samples after 1 h exceeded the cumulative concentrations in the cooler soil samples. Of the total applied MeBr, 32%, 35%, and 54% volatilized in the 15° C, 25° C, and 35° C samples, respectively, over 5 d of the experiment. Over 85% of the total MeBr flux occurred within the first 3 h after fumigation at all soil temperatures tested.

Table 1. Volatility of methyl bromide (MeBr), as influenced by soil temperature, expressed as a percentage of the total MeBr initially applied

Soil Temperature (° C)	Soil Moisture (bar)	% Volatilized (3 h)	Total % Volatilized (119 h)
15°	0.3	27.3[a]	32.2[a]
25°	0.3	30.4[a]	35.2[a]
35°	0.3	50.9[b]	54.4[b]

Values followed by the same letter are not significantly different ($p \geq 0.05$)

Volatility of MeBr significantly increased at higher soil moisture (Table 2). The quantities of MeBr that volatilized from the 0.3 and 1 bar soils were not significantly different. Of the total applied MeBr, 4%, 29%, 35%, and 67 % volatilized from the 3, 1, 0.3, and 0.03 bar samples, respectively. Most of the MeBr flux occurred within 2 h after fumigation at all soil moistures tested. Our results are consistent with previous research that shows that low soil moisture leads to increased adsorption of MeBr. Chisholm and Koblitsky (20) also observed a greater adsorption of MeBr in dry soils than wet soils. Although MeBr is believed to be weakly adsorbed, volatility may have increased as a result of competition between water and MeBr molecules for the same sorption sites.

Table 2. Volatility of methyl bromide (MeBr), as influenced by soil moisture, expressed as a percentage of the total MeBr initially applied

Soil Moisture (bar)	Soil Temperature (° C)	% Volatilized (2 h)	Total % Volatilized (72 h)
3	25°	4.0[a]	4.1[a]
1	25°	28.7[b]	28.9[b]
0.3	25°	28.0[b]	34.7[c]
0.03	25°	66.3[c]	66.7[d]

Values followed by the same letter are not significantly different ($p \geq 0.05$)

Degradation Studies. Bromide ion was measured to determine the influence of temperature and moisture on the degradation of MeBr in fumigated soil samples. MeBr degradation significantly increased at higher temperatures (Table 3). Samples incubated at 35° C contained from 7-20x more Br⁻ than soils at 25° C and 15° C, respectively. After 211 h, 1.1, 4.6, and 22.6% of applied MeBr had degraded to Br⁻ in the 15° C, 25° C, and 35° C soil samples, respectively. From this study it was not clear whether the transformation of MeBr was abiotic, biotic, or a combination of the two. Gentile et al. (13) reported a decrease in MeBr half-life in static-anaerobic water samples at higher temperature and pH.

Table 3. Transformation of methyl bromide (MeBr), as influenced by soil moisture and soil temperature, expressed as a percentage of the total MeBr initially applied

Soil Temperature (° C)	Soil Moisture (bar)	% Degraded	Time (post fumigation)
25°	0.03	3.4[a]	72 h
25°	0.3	3.6[a]	72 h
25°	1	1.0[b]	72 h
25°	3	1.5[c]	72 h
15°	0.3	1.1[b]	119 h
25°	0.3	4.6[a]	119 h
35°	0.3	22.6[d]	119 h

Values followed by the same letter are not significantly different (p ≥ 0.05)

MeBr degradation was also increased significantly at the highest soil moistures (0.03 bar and 0.3 bar) compared to the two driest soils (1 bar and 3 bar) tested (Table 3). However, only minimal amounts (≤ 3.6%) of MeBr were degraded even at the highest soil moisture. Fumigated soil samples with moisture levels above field capacity contained > 2x more Br⁻. MeBr is known to be hydrolyzed to Br⁻ and methanol in water (13). Yagi et al. (4,21) compared two field fumigation experiments that measured the flux of MeBr and formation of Br⁻ in soil. They observed an increase in Br⁻ and a decrease in MeBr volatility related to a combination of increased soil moisture, pH, organic matter, and injection depth.

Microbial Toxicity. Respiration was measured in soils fumigated with MeBr to determine the effect on soil microorganisms. A typical structural fumigation rate (594 g/m³) and 2.5x a typical field fumigation rate (350 μg/g) were used. Soil samples fumigated at 594 g/m³ (structural rate) sustained depressed respiration throughout the 24-d experiment compared to control (unfumigated) soils. The 350 μg/g soil treatment caused a temporary depression in CO_2 efflux, but it was not significantly different from the control (unfumigated) soil after 4 days. A reduction in soil respiration suggests a reduction in microbial biomass and/or activity. MeBr is a broad-spectrum, nonselective fumigant that kills soil-borne pathogens as well as beneficial microorganisms. Sensitivity to MeBr varies; however, all organisms would be expected to be susceptible to high concentrations. At the 350 mg/g treatment (2.5x a typical field rate), the microbial population appeared to be able to recover from the MeBr-induced toxicity. A similar or increased recovery might be expected in soils fumigated at the field rate (392 kg/ha = 132 µg MeBr/g soil).

Soil Column Study. Undisturbed soil columns were used to study the volatility, degradation, movement, and leaching potential of MeBr. MeBr volatilized rapidly from the soil column. Headspace analysis of the column indicated that MeBr flux peaked at 32 h after fumigation, with no detection after 21 d. These results are similar to the results of the laboratory studies and to those obtained by Yagi et al. (4) in field studies of MeBr. Namely, MeBr is rapidly volatilized from fumigated soil.

Soil column leachates from each rain event were analyzed for MeBr and Br⁻. MeBr was not detected in any of the leachate samples during the experimental period (23 weeks). Br⁻ concentration increased from a background levels (0.004 mg) to 6.45 mg within the first rain event following fumigation (Figure 2). Levels of Br⁻ continued to increase, peaked at 3 weeks (68.4 mg), and gradually decreased with subsequent rain events. A total of 460 mg Br⁻ leached through the soil column, which represents > 5% of the MeBr initially applied. Wegmand et al. (14) detected MeBr and Br⁻ in drainage water from fumigated glasshouse soils. In addition, they observed a sharp increase in Br⁻ concentration during initial irrigation of the soils, followed by a steady decrease. The absence of MeBr in the leachate in the current study indicated MeBr did not leach through the soil profile of the undisturbed soil columns.

After 23 weeks the soil column was divided into 5-cm fractions and analyzed for Br⁻. Levels of Br⁻ were similar to control (untreated) soil samples. The increased quantity of Br⁻ in the leachate and no detection of residual MeBr and Br⁻ in the soil profile at the completion of the test, imply that MeBr degraded in the soil.

Field Study. The field fumigation study showed that 43% of the applied MeBr was volatilized within 4 d (Figure 3). In addition, 18% of the applied MeBr escaped through the plastic tarp during the first 48 h. A sharp rise in MeBr flux occurred following the removal of the tarp; an additional 24% of the applied MeBr volatilized from the soil within the next 24 h. Only trace amounts of MeBr were detected after day 4 of the experiment. Yagi et al. [4,21] reported a 34% and 87% flux of MeBr within 7 days from fumigated fields, depending on soil moisture and other soil properties.

Concentrations of MeBr in soil gas were also measured at various time intervals. At 48 h post fumigation, 8% of the applied MeBr was detected in the soil gas. MeBr concentrations in the soil gas dissipated rapidly with time; only trace amounts of MeBr were observed after 10 d. Yagi et al. [21] reported negligible quantities of MeBr in the soil gas after 7 d.

Soil samples from the fumigated fields were also analyzed for Br⁻ as an indicator of MeBr transformation. Concentrations of Br⁻ in soil after MeBr fumigation were significantly greater ($p \leq 0.05$) than in the control soil samples collected prior to fumigation. Based on a measured increase in Br⁻ concentration in the soil, a significant portion of the applied MeBr (30%) had degraded within the first 2 days. Levels of Br⁻ decreased with time and returned to background (pre-fumigation) level within 24 d.

Microbial respiration was measured to determine the potential toxicity of MeBr applied at the field rate (390 kg/ha) to soil microorganisms. MeBr applied at the field rate was apparently not toxic to the microbial population. There was no significant difference ($p \leq 0.05$) in microbial respiration between the control (unfumigated) and fumigated soil samples. Previously, we performed similar studies using quantities of MeBr that represented a typical structural fumigation rate and 2.6x typical field application rate. Sustained depression in microbial activity was observed only in soils fumigated at the structural fumigation rate.

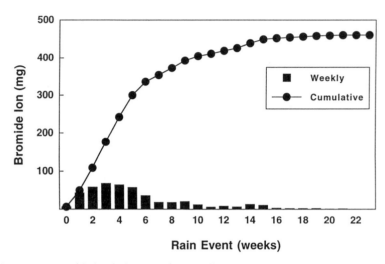

Figure 2.   Bromide ion in leachate from undisturbed soil columns fumigated with methyl bromide.

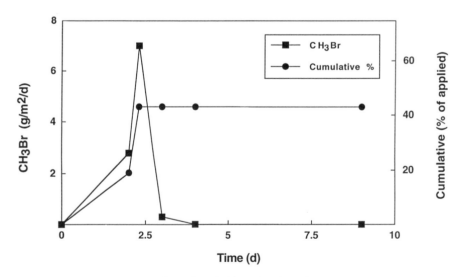

Figure 3.   Volatility of methyl bromide from fumigated fields.

Conclusions

The influence of soil environmental variables on MeBr was studied to increase our understanding of MeBr transformation and movement in soil. Our research showed that differences in temperature and soil moisture significantly influence MeBr fate. At higher temperature and soil moisture, the volatility of MeBr increased. The majority of the MeBr flux from the soil occurs within a few hours after the initial fumigation. A similar trend was observed with the degradation of MeBr in the soil. MeBr degraded more rapidly at the higher soil temperatures and moistures; however, only at the highest temperature tested (35°) did a significant portion of the applied MeBr get transformed. Studies with undisturbed soil columns and field studies both confirmed the laboratory studies on MeBr volatility and transformation. MeBr does induce toxicity upon fumigation into soil. The concentration of MeBr applied to the soil determines whether or not microbial communities are able to recover from the chemically induced toxicity.

Acknowledgment

This research was supported by a grant from the National Agricultural Pesticide Impact Assessment Program (NAPIAP). We express our thanks to Great Lakes Chemical Co. for supplying the methyl bromide used in this study and to Pam Rice, Theresa Klubertanz, and Mark Petersen for help with soil column collection. Journal paper No. J-16595 of the Iowa Agricultural and Home Economics Experiment Station Project No. 3187.

Literature Cited

1.   Strub, D. In *Into the Sunlight: Exposing Methyl Bromide's Threat to the Ozone Layer*; Strub, D., Malakoff, D., Eds.; Public Interest Publications: Washington, DC, **1992**; pp 2-10.

2.   Singh, H. B.; Kanakidou, M. *Geophys. Res. Lett.* **1993**, 20, 133-136.

3.   Khalil, M. A. K.; Rasmussen, R. A.; Gunawardena, R. *J. Geophys. Res.* **1993**, 98, 2887-2896.

4.   Yagi, K.; Williams, J.; Wang, N.-Y.; Cicerone, R. J. *Proc. Nat. Acad. Sci.* **1993**, 90, 8420-8423.

5.   Wofsy, S. C.; McElroy, M. B.; Young, Y. L. *Geophys. Res. Lett.* **1975**, 2, 215-218.

6.   National Agricultural Pesticide Impact Assessment Program (NAPIAP) United States Department of Agriculture In The Biologic and Economic Assessment of Methyl Bromide; NAPIAP: Washington, DC, **1993**; pp 1-7.

7.   Zurer, P. S. *Chem. Eng. News* **1993**, February, 23-24.

8.   Howard, P. H. In *Handbook of Environmental Fate and Exposure Data*; Howard, P. H., Ed.; Lewis Publishers: Chelsea, MI, **1989**; pp 386-393.

9.   Oremland, R. S.; Miller, L. G.; Strohmaler, F. E. *Environ. Sci. Technol.* **1994**, 28, 514-520.

10.   Rasche, M. E.; Hyman, M. R.; Arp, D. J. *Appl. Environ. Microbiol.* **1990**, 56, 2568-2571.

11.   Vogel, T. M.; Criddle, C. S.; McCarty, P. L. *Environ. Sci. Technol.*
      1987, 21, 722-737.
12.   Shorter, J. H.; Kolb, C. E.; Crill, P. N.; Kerwin, R. A.; Talbot, R.
      W.; Hines, M. E. *Nature.* **1995**, 377, 717-719.
13.   Gentile, I. A.; Ferraris, L.; Crespi, S. *Pestic. Sci.* **1989**, 25, 261-272.
14.   Wegman, R. C. C.; Greve, P. A.; De Heer, H.; Hamaker, P. *Water
      Air Soil Pollut.* **1981**, 16, 3-11.
15.   Woodrow, J. E.; McChesney, M. M.; Seiber, J. N. *Anal. Chem.*
      1988, 60, 509-512.
16.   Riga, T. J.; Lewis, E. T. *Am. Environ. Lab.* **1995**, 4, 14-15.
17.   Edwards, N. T. *Soil Sci. Soc. Am. J.* **1982**, 46, 1114-1116.
18.   Singh, P.; Kanwar, R. S. *J. Environ. Qual.* **1991**, 20, 295-300.
19.   Kruger, E. L.; Somasundaram, L.; Kanwar, R. S.; Coats, J. R.
      Environ. Toxicol. Chem. **1993**, 12, 1969-1975.
20.   Chisholm, R. D.; Koblitsky, L. *J. Econ. Entomol.* **1943**, 36, 549-
      551.
21.   Yagi, K.; Williams, J.; Wang, N. -Y.; Cicerone, R. J. *Science* **1995**,
      267, 1979-1981.

Chapter 6

Degradation of Methyl Bromide by Green Plants

Peter M. Jeffers[1] and N. Lee Wolfe[2]

[1]Chemistry Department, Bowers Hall, P.O. Box 2000,
State University of New York, Cortland, NY 13045
[2]Ecosystems Research Division, National Exposure Research Laboratory,
U.S. Environmental Protection Agency, 960 College Station Road,
Athens, GA 30605–2700

Eleven experiments with five different green plants showed absorption and/or degradation of methyl bromide both in aqueous solution and in the gas phase. *Myriophyllum aquatica* (parrot feather), *Iris pseudocorus* (swamp iris), *Draparnaldia* and *Spirogyra* (two filamentous fresh-water algae), and *Oxalis corniculata* (trefoil, or creeping wood-sorrel) removed methyl bromide with a half-life of 10 to 36 hours for initial gas concentrations from 5-20 ppm (volume) and a mass ratio of air to plant within the range of 0.03 to 50. Reactive removal of methyl bromide from the atmosphere by green plants may be an important pathway to incorporate into global balance schemes that aim to determine the significance of the anthropogenic load of the chemical to the global total.

Methyl bromide is a major carrier of bromine atoms to the stratosphere, and the 50-fold higher efficiency of bromine compared to chlorine in the ozone destruction chain makes the relatively small amount of atmospheric methyl bromide an important contributor to total stratospheric ozone loss. The major quantified tropospheric sink is reaction with hydroxyl radicals, while the oceans have been cited as both a source and a reactive sink. Clearly, atmospheric methyl bromide originates from both natural and anthropogenic sources. The chemical is biosynthesized in the oceans; its oceanic abundance appears to correlate with oceanic chlorophyll (1), and it is liberated to the atmosphere during biomass burning. Butler (2) and Yvon and Butler (3) have incorporated the major known formation and destruction pathways into a model that allows determination of the critical elements in establishing a global balance scheme for methyl bromide. A proper evaluation of the significance of anthropogenic methyl bromide depends on identification and careful measurement of all the important degradation mechanisms. We have measured and reported the rate of abiotic hydrolysis and loss by chloride ion exchange in sea water (4), but in this paper we suggest a new sink whereby methyl bromide is metabolized or irreversibly adsorbed by green plants. Metabolism of methyl bromide has been observed previously. Oremland *et al.* (5) report degradation in anaerobic sediments due to sulfide

0097–6156/96/0652–0053$15.00/0

attack on methyl bromide, with the sulfide produced by methanogenic bacteria. Shorter *et al.* (6) observed rapid uptake of methyl bromide by a variety of soils, and have suggested that this soil sink may reduce the atmospheric lifetime by 50%. Our suggestion that green plants may be an additional sink for methyl bromide is supported by previous work in this lab and elsewhere (7) wherein plant dehalogenase enzymes capable of rapidly dehalogenating a wide range of chlorinated, brominated, and iodinated hydrocarbons were isolated.

Experimental

Myriophyllum aquatica (Parrot Feather). Stems and the feathery leaves totaling 2.0 g of the aquatic plant parrot feather (0.26 g dry wt.) were placed in a 500-ml Erlenmeyer flask fitted with a foil-faced rubber stopper. A 10-cm syringe needle extended through the stopper into the center of the flask. The hub of the needle had been plugged with a septum, through which gas samples were withdrawn for analysis, using a 0.1 ml gas-tight syringe. A matching 500-ml flask was prepared without any plant material, to serve as a control. Both flasks were spiked with gaseous methyl bromide to give a starting concentration of about 5 ppm by volume. Thorough mixing was effected by shaking and inverting the flasks periodically. The methyl bromide concentrations were monitored by GC analysis (ECD detection, 100 µl gas sample) for 48 hours. Another experiment utilized 2.25 g parrot feather and a methyl bromide dosing of 200 ppm. This system was checked for methane production (GC with FID detection capable of measuring methane at normal ambient air concentrations) at times from 81 to 145 hr, and bromide ion was measured in ground plant tissue extracted with deionized water at 145 hr (Dionex ion chromatograph).

Iris pseudocorus (Swamp Iris). An entire swamp iris plant weighing 18.5 g was placed in a 500-ml flask, as described above, with 25 ml of swamp water. A control flask also contained 25 ml of water from the plant's native site. Both flasks were injected with gaseous methyl bromide to an initial gas phase concentration of about 20 ppm by volume. Methyl bromide concentrations were monitored for 114 hours, after which the iris plant was transferred, dry, to a clean 500-ml flask, from which samples were taken for 24 hours. After sampling was completed, the plant was removed and placed in a jar of swamp water. The plant remained in apparent good health. The water in contact with the plant and the water in the control flask were sampled at 114 hours, when the initial phase of the experiment concluded.

Fresh water algae. Algae were harvested from a local pond, and microscopic examination indicated that it contained a few strands of *Spirogyra* but was in bulk filamentous *Draparnaldia*, with healthy co-populations of assorted microbes. A lightly drained sample of 28 g of the algae was placed in a 500 ml flask, as above, with a control flask prepared using 28 g of pond water decanted from the algae. Each flask was injected with gaseous methyl bromide at an initial gas-phase concentration of 20 ppm by volume.

A second experiment with *Draparnaldia* was performed using 28.9 g of the lightly drained algae with 29.1 g of water poured off the algae and filtered through a "coarse analytical" gravity-feed funnel filter paper serving as the control. Each system contained numerous microorganisms, but no plant filaments were observed in the filtered water.

A third experiment with algae utilized 9.1 g "drained" *Draparnaldia* (0.05 g dry weight of the algae determined by filtering and drying at the conclusion of the run), 8.1 g *Spirogyra* (0.10 g dry plant weight), and 9.3 g of water filtered off the *Spirogyra*, as above, serving as the control.

Oxaliscorniculata (trefoil, or creeping wood-sorrel) was gathered from the lawn outside the laboratory. A total of 100 g of the plants, with rinsed roots, and 25 ml of deionized water was placed in five 50 ml beakers evenly distributed along a 1 m x 10 cm x 10 cm glass reactor. The seams of the reactor were sealed with epoxy cement, and the seal of the reactor to the bottom plate was made with Apiezon Q black wax. Spiking and sampling were done through Teflon-faced septa waxed over five 5 mm holes spaced evenly along the top plate. Methyl bromide was added at an initial concentration of 5 ppm. A matching control reactor was prepared and treated exactly the same, except that no plants were placed in the beakers of water. The plant reactor was re-spiked with methyl bromide at 89 hrs, and again at 148 hr, and methyl bromide concentrations in both plant and control reactors were monitored for a total of 196 hr. Uniformity of mixing was checked by extracting samples from several sample ports at each time interval. These plants were maintained under a grow light with a 16 hour duty cycle. All experiments we discuss in this paper were performed at the constant laboratory temperature of 22 °C. We cannot discount the rhizosphere effects within this system, but the packing of the plants into the beakers presented far more leaf surface to the gas within the reactor and a difficult path to the water surface and the root area.

Results and Discussion

The studies we report were predicated on work done in this laboratory on degradation of aqueous halogenated hydrocarbon solutions by sediments and by aquatic plants and freshwater algae that contain dehalogenase enzymes (7). The parrot feather and iris plants used in this study have both been shown to contain dehalogenase enzymes, but previous tests for activity toward a gas-phase halocarbon have not been published.

Methyl bromide concentration decreased from 5 ppm by about a factor of 4 in 70 hr with the parrot feather (Figure 1). Leafy plant pieces had been severed from the root systems for this experiment, and by 70 hours the plant material no longer appeared vital. The concentration of methyl bromide followed a logarithmic decay, with a first-order decay constant of about 0.02 hr^{-1}. This 35-hour half-life was achieved with a mass-ratio of air to plant of about 0.27 (500 ml air to 2 g of parrot feather). Methyl bromide concentration in the control flask decreased by less than a factor of 2 during the same experimental period. With each sampling, about 0.4 ml of gas was withdrawn in order to clear the volume of the sampling needle, which accounts for some of the observed loss from the control flask, but there was additional loss past the foil- faced stopper. (More recent experiments with foil and septum sealed screw-cap test tubes indicated that loss could be rendered negligible in the control systems.) Product analysis with the parrot feather systems yielded some important information. Our analysis for methane was sufficiently sensitive to see the correct methane concentration in ambient air (on the order of 1 ppm). Analysis of the methyl bromide/parrot feather system showed no additional methane production, although a significant amount of methyl bromide had disappeared. However, we did find bromide ion in the aqueous extract of ground plant tissue. Bromide was the only anion to increase in concentration over the blank system, and the amount found accounted for all the methyl bromide delivered to the flask.

METHYL BROMIDE / PARROT FEATHER

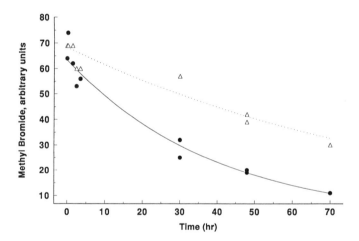

Figure 1. Loss of methyl bromide from 500 ml flasks. Triangles are the control system, circles the flask containing 2 g of *Myriophyllum aquatica*.

With the iris plant, the methyl bromide concentration decreased by a factor of nearly 20 within 114 hr, from an initial concentration of 20 ppm. This decrease, treated as first-order decay, had a rate constant of about 0.025 hr^{-1}, or a half-life of about 28 hours, and was achieved with an air to plant mass ratio of about 0.03 (500 ml air to 18 g plant). Concentration of methyl bromide in the control flask decreased by less than 30% during this experiment. After 114 hr, the iris was placed in a clean, dry 500-ml flask. Air from this flask was checked for the next 24 hours, and no methyl bromide was detected in the vapor phase. A check of the water in both the plant-containing flask and the control flask showed methyl bromide concentrations at the values calculated from the Henry's law constant and the final gas-phase concentrations. With the volumes of our system, over 80% of the methyl bromide resided in the vapor phase.

Methyl bromide concentrations dropped by a factor of four within 48 hours with the first sample of algae and also in the control flask which contained pond water decanted from the algae. Microscopic examination (200x magnification) disclosed active microbe populations in each flask, and numerous algae filaments in the decanted water. A second experiment was performed using 30 g of decanted water in one 500 ml flask and 30 g of decanted water filtered through a coarse analytical filter paper by gravity in the control. Microscopic examination showed active microbes in each flask at about equal numbers, but no plant filaments in the filtered water. Methyl bromide concentration decreased in the decanted water flask at a rate comparable to the loss observed with the algae and the previous decanted water control flask, but the loss rate was 50% smaller in the filtered water flask.

In our final experiment with *Draparnaldia* and *Spirogyra* we varied the amount of plant and the initial CH_3Br, and we used, once again, filtered water as the control. In all three systems the CH_3Br concentration decreased exponentially, with a smaller relative change in the control system, but the amount of reaction had little correlation to the starting concentration or to the amount of dry plant material. A check with ion chromatography at the end of the experiment showed that bromide ion had been produced in the algae systems. The lack of correlation with amount of plant matter suggests that in the algae systems the rate of methyl bromide reaction is limited by transfer from the gas to the liquid phase, and that the reaction of dissolved CH_3Br with plant dehalogenase enzymes is sufficiently rapid that any amount of viable plant matter present will suffice, within the limits of our experimental design.

Figure 2 shows the entire course of the multiple spike methyl bromide - trefoil system. The control reactor had a first-order loss/leak rate corresponding to a half-life of 54 hr. Loss of methyl bromide in the trefoil containing reactor was approximately six times as rapid, and showed an acceleration at reduced concentrations. The second and third spikes resulted in more rapid loss than that observed with the initial introduction of methyl bromide. Both these behaviors can be related to expected enzyme kinetics behavior. The plant enzyme responsible for dehalogenation appears to be a senescence enzyme (8), one that is expected to increase in concentration as the plant is stressed, thus the second and third spikings should be degraded more rapidly than the first. The low concentration apparent rate increase is consistent with Michaelis-Menten enzyme kinetics, as illustrated by Figure 3 where an enzyme kinetics simulation provides a better fit for the last two points than does a first-order decay curve.

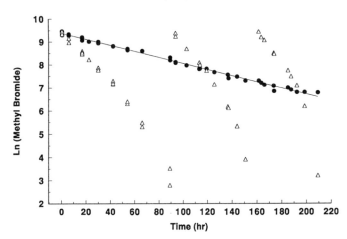

Figure 2. Loss of methyl bromide from the 10-1 reactors. Circles - control system, triangles - *Oxalis corniculata*. The plant system was re-spiked at 90 and at 160 hr. Note the change from first-order decay at low methyl bromide concentrations. Initial methyl bromide concentrations were all 5 ppm (volume).

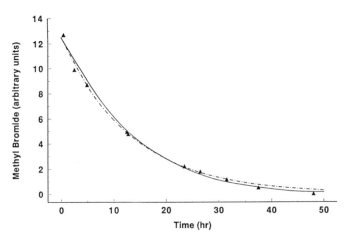

Figure 3. Details of the third spike of the *Oxalis corniculata* system from Figure 2, showing both a first-order fit (dashed line) and a Michaelis-Menten fit (solid line), with fitting parameters $C_o = 12.5$, $V_M = 2.0$, $K_M = 20$. The enzyme kinetics model is a better fit of the last two points which clearly diverge from first-order decay (see Figure 2).

These experiments strongly indicate that methyl bromide was metabolized by the plants, and they do indicate a significant loss on a time scale that is short relative to hydrolysis or chloride exchange. This loss represents a process that can occur in the canopy of green plants. Previous experiments (9) have shown chemical reduction of aqueous halogenated hydrocarbons by plant dehalogenase enzymes; thus, a reactive mechanism in the gas phase should not be surprising. Further, more quantitative experiments involving additional species of plants and covering methyl bromide concentrations closer to ambient global atmospheric levels certainly are warranted. Our own related experiments with gas phase haxachloroethane degradation to tetrachloroethylene by green plants (work in progress) indicate that the degradation process scales with the leaf area but also varies from species to species. Proper factoring-in of oceanic degradation of methyl bromide and atmospheric reactive loss in the green plant canopy as well as by oceanic algae might have a significant influence on determining the relative importance of the anthropogenic load of methyl bromide on the total global balance. It appears possible that the combined buffering capacity of soils and green plants may render the release of methyl bromide during agricultural use insignificant in changing the net atmospheric load.

Acknowledgments. This work was performed while PMJ was a National Research Council Senior Fellow at the USEPA Laboratory, Athens, GA. The research was funded by the Strategic Environmental Research and Development Program (SERDP).

Literature Cited

(1) Anbar, A.D.; Yung, Y.L.; Chavez, F. P. *Global Biochemical Cycles,* **1996.**

(2) Butler, J.H. *Geophys. Res. Lett.* **1994,** *21,* 185.

(3) Yvon, S. A.; Butler, J. H. *Geophys. Res. Lett.* **1996,** *23,* 53.

(4) Jeffers, P. M.; Wolfe, N. L. *Geophys. Res. Lett.* **1996,** *23.*

(5) Oremland, R. S.; Miller, L. G.; Strohmaler, F. E. *Environ. Sci. Technol.* **1994,** *28,* 514.

(6) Shorter, J. H.; Kolb, C. E.; Crill, P. M.; Kerwin, R. A.; Talbot, R. W.; Hines, M. E.; Harriss, R. C. *Nature,* **1995,** *377,* 717.

(7) Schnoor, J.L.; Licht, L. A.; McCutcheon, S. C.; Wolfe, N. L.; Carreira, L. H. *Environ. Sci. Technol.* **1995,** *29,* 318A

(8) Carreira, L. H., Personal communication.

(9) Nzengung, V. A.; Wolfe, N. L.; Carriera, L. H.; McCutcheon, S. C. *Environ. Sci. Technol.* **1996,** (submitted for review).

Chapter 7

Fate of 1,3-Dichloropropene in Aerobic Soils

F. R. Batzer, J. L. Balcer, J. R. Peterson, and J. D. Wolt

Global Environmental Chemistry Laboratory–Indianapolis Laboratory, DowElanco, 9330 Zionsville Road, Indianapolis, IN 46268–1054

The degradation of the soil fumigant, 1,3-dichloropropene (1,3-D), was investigated to determine the rate of degradation of 1,3-D and the identity of metabolites in aerobic soils. Studies were conducted in the dark at 25 °C with uniformly ^{14}C-labeled 1,3-D at a concentration of approximately 100 μg/g on three soils: Wahiawa silty clay, Catlin silt loam and Fuquay loamy sand. Aerobic soil half-lives for 1,3-D were 1.8, 12.3, and 61 days on the Wahiawa silty clay, Catlin silt loam, and Fuquay loamy sand, respectively.

Degradation of 1,3-D resulted in the formation of 3-chloroallyl alcohol, 3-chloroacrylic acid, numerous minor carboxylic acid metabolites, and carbon dioxide. In addition, there was also extensive incorporation of ^{14}C labeled material into the soil organic matter of all soils.

The chemical 1,3-dichloropropene has been widely used as a soil fumigant and is currently registered as an active ingredient in Telone[*] brand soil fumigants. It has been used to treat fields intended for vegetables and field crops such as cotton, potatoes, sugar beets, tobacco, and pineapples. Typical application rates for Telone II soil fumigant for field crop use on mineral soils range from 130 to 195 kg/ha with 388 kg/ha being the maximum rate (1).

The environmental fate of 1,3-dichloropropene (1,3-D) has been studied extensively under laboratory and field conditions. Laboratory studies involving 1,3-D have provided information on hydrolysis rates (2), aerobic soil degradation rates (3, 4, 5, 6, 7, 8, 9, 10, 11), anaerobic aquatic degradation rates (12), photochemical degradation rates in air (13), and sorption coefficients (14, 15). Field studies have provided information on field dissipation (8), field volatility (16), and ground water monitoring (17, 18) for 1,3-D. These studies indicate that environmental dissipation

---

[*] Trademark of DowElanco.

0097–6156/96/0652–0060$15.00/0
© 1996 American Chemical Society

of 1,3-D occurs by four major routes: volatilization, hydrolysis, metabolism and tropospheric reactions with hydroxyl radicals.

Hydrolysis of 1,3-D is a major pathway for degradation (2) which is independent of pH. The half-life of 1,3-D was 11 days in sterile buffer (pH 7) at 20 °C (2). The major product of 1,3-D hydrolysis was identified as 3-chloroallyl alcohol (*cis/trans*-3-chloroprop-2-en-1-ol).

The degradation of 1,3-D on aerobic soils has been examined by numerous investigators (3 through 11). Half-lives for 1,3-D varied from a low of 1.8 days on a silty clay soil at 25 °C (7) to about 6 days on clay soils to 17 days on sandy soils at 20 °C (4). Research from several investigators has shown that abiotic and biotic aerobic soil degradation of 1,3-D occurs (5, 10, 11). In some cases, the effect of repeated application is shorter 1,3-D half-lives on aerobic soils (10, 11). Major metabolites of 1,3-D have been identified as 3-chloroallyl alcohol, 3-chloroacrylic acid, and carbon dioxide (6). Roberts and Stoydin (6) had also detected other polar metabolites which were not identified. In addition, there was extensive incorporation of ^{14}C material into soil organic matter.

The microbial metabolism of 3-chloroallyl alcohol is rapid in both topsoils and subsoils (3,9). Half-lives of 1.2 to 3.1 and 0.4 to 1.1 days for *cis*- and *trans*-3-chloroallyl alcohol, respectively, were observed in clay topsoils (3). In another study, degradation of 3-chloroallyl alcohol was also rapid in topsoils (9) with half-lives on the order of 2 to 4 days for the *cis*-isomer and 1 to 2 days for the *trans*-isomer. Roberts and Stoydin (6) reported that 3-chloroallyl alcohol was rapidly transformed into 3-chloroacrylic acid (*cis/trans*-3-chloroprop-2-enoic acid) in a loam soil but that in a sandy loam soil 3-chloroallyl alcohol was still found in significant amounts (>25% of applied) after 12 weeks. Microbial breakdown of 3-chloroallyl alcohol to 3-chloroacrylic acid in cultures of *Pseudomonas* species isolated from soil was reported by Belser and Castro (19). Further metabolism of 3-chloroacrylic acid was also observed (6, 19, 20).

Purpose.

The fate of 1,3-D in aerobic soils from the USA was investigated to determine the rate of 1,3-D degradation in top soils and subsoils. A major goal of the research was to develop an incubation system that could maintain material balance for a volatile organic compound and not capture CO_2 during the incubation. Another goal was confirmation of 3-chloroallyl alcohol and 3-chloroacrylic acid as aerobic soil degradates. Because previous work (6) had indicated that additional metabolites arose from the aerobic soil metabolism of 1,3-D and its primary metabolites, identification efforts were conducted on the minor metabolites that were produced. These results were generated and reported as part of a re-registration submission in the United States to support the use of 1,3-D as a soil fumigant.

Materials and Methods.

Soils. The following soils were used: Fuquay loamy sand, Catlin silt loam, and Wahiawa silty clay Horizon A and Horizon B. After receiving the fresh soils, they were sieved through a 2 mm screen and stored at approximately 4 °C. Soils were analyzed for particle size distribution, organic carbon content, pH, cation exchange capacity, and soil moisture content (Table I).

Test Materials and Dosing Solutions. Dosing solutions were prepared from stock solutions of ^{14}C-radiolabeled and unlabeled 1,3-D in acetone. The nonlabeled material consisted of a 50/50 mixture of *cis*-1,3-D (93.8%) and *trans*-1,3-D (96.5%). The final specific activity of dosing solutions ranged from 0.08 mCi/mmole to 1.15 mCi/mmole. The ^{14}C radioactivity of dosing solutions was quantitated by liquid scintillation counting (LSC). The ^{14}C radiochemical purity of dosing solutions was determined by HPLC to be 99%.

Other Radiolabeled Standards and Reagents. Metabolites of 1,3-D were obtained as ^{14}C-radiolabeled standards: 3-chloroallyl alcohol (specific activity 5.4 mCi/mmole, radiochemical purity 99+%), and 3-chloroacrylic acid (specific activity 3.8 mCi/mmole, radiochemical purity 98.5%). Additional ^{14}C-radiolabeled carboxylic acids were purchased from Sigma Chemical Company as standards for HPLC analyses and as possible metabolite standards (all radiochemical purities 98% or greater). These standards included acetic acid sodium salt (specific activity 48.9 mCi/mmole), citric acid (specific activity 60.0 mCi/mmole), formic acid sodium salt (specific activity 57.0 mCi/mmole), oxalic acid (specific activity 4.5 mCi/mmole), and propionic acid sodium salt (specific activity 4.6 mCi/mmole). In addition, ^{14}C-labeled butanol (Sigma, specific activity 3.1 mCi/mmole, radiochemical purity 98%) was obtained and used to prepare ^{14}C-labeled esters from carboxylic acid standards.

The following acids were purchased from Aldrich Chemical Company unless otherwise noted and were used to prepare their respective butyl esters: acetic acid (Fisher), adipic acid, butyric acid, chloroacetate, *cis*-3-chloroacrylic acid, *trans*-3-chloroacrylic acid, 4-chlorobutyric acid, fumaric acid, glycolic acid, hexanoic acid (Sigma), lactic acid, malic acid, malonic acid, 2-methylmalonic acid, oxalic acid, propionic acid, and succinic acid.

High Performance Liquid Chromatography (HPLC) grade solvents and reagents were obtained from either Fisher Scientific or Aldrich Chemical Company. Water was obtained from a Corning Mega-Pure® Glass Water Still or Fisher Scientific (HPLC grade). Ultima Gold™ scintillation cocktail and Permafluor® V scintillation cocktail were obtained from Packard Instrument Company. For combustion recovery determinations, Harvey [14]C Cocktail from R. J. Harvey Instrument Corporation (Hillsdale, NJ) was used.

HPLC Methods. High Performance Liquid Chromatography (HPLC) was conducted with Waters™ HPLC systems (Millipore Corporation, Milford, MA) consisting of either Model 45 pumps, Model 720 System Controller, Model 740 Data Module, a Kratos Spectroflow Model 757 LC Spectrophotometer, and an Isco Retriever® IV fraction collector or a Waters Model 600E pump and Controller system with an Isco Retriever® IV fraction collector. Solvent A was 1% acetic acid in water and Solvent B was 1% acetic acid in acetonitrile. A Waters Nova-Pak® Radial-Pak C_{18} column was used for general HPLC analyses. Gradient 1 was used to analyze acetone extracts from carbon traps and from the soil samples and was comprised of the following steps: 1) 100% Solvent A for 20 minutes at 1 mL/min, 2) linear ramp over 5 minutes to 100% Solvent B, 3) 100% Solvent B for 5 minutes at 2 mL/min, 4) linear ramp to 100% Solvent A over 5 minutes at 2 mL/min and hold for 10 minutes to re-equilibrate. Gradient 2 was used to analyze solutions containing the butyl esters of metabolites and standards at a flow rate of 1.5 mL/min and consisted of the following steps: 1) Step ramp from 100% Solvent A to 60% Solvent A, 2) linear ramp from 60% to 0% Solvent A, 3) hold 100% Solvent B for 10 minutes, 4) ramp to 100% Solvent A over 1 minute, 5) hold at 100% Solvent A for 10 minutes to re-equilibrate. Ion exclusion HPLC was conducted with an Interaction® ION-300 HPLC column (0.78 x 30 cm) with 0.05 N sulfuric acid as eluent at a flow rate of 0.4 mL/minute.

Radioanalysis (Liquid Scintillation Counting) and Radioactivity Standards. Measurements of radioactivity were made using a Packard Liquid Scintillation Spectrometer Model 2250CA or Model 2500TR. Samples were counted for at least 3 minutes or to a 2 sigma (95%) confidence level. The internal quench curves for the liquid scintillation counters (LSC) were obtained by counting a set of quenched [14]C

LSC sealed standards. Reference ^{14}C sealed standards purchased from Packard Instrument Company (Downers Grove, IL) were used throughout the study to maintain a calibration record for each of the liquid scintillation counters.

Incubation Flask Design. Due to the volatility of 1,3-D (vapor pressure 34.3 mm Hg (21) and 23.0 mm Hg (22), for the *cis*- and *trans*-1,3-D, respectively) an incubation flask was designed to contain 1,3-D, volatile metabolites, and CO_2. A caustic trap could not be used due to the rapid hydrolysis of 1,3-D at 25 °C in aqueous solution. Therefore, a sealed, static incubation flask was designed which could be accessed without opening for treatment and sampling. As a sealed system, the replenishment of oxygen is not possible during incubation.

The system was based on a standard 250-mL Erlenmeyer flask with three modifications (Figure 1). The flask neck consisted of a 2.5 cm outer diameter (o.d.) heavy-wall glass tubing. A small, threaded side-arm (1-mL Reacti-Vial™ (Pierce 13221)) was added at the side of the flask which, when equipped with a Mininert^R screw cap, afforded a Teflon^R-sealed access port to the flask. Another side-arm for purging was added to the opposite side of the flask using 12 mm (o.d.) glass tubing. The end of this tube was placed in the center of the flask approximately 2.5 cm from the inside bottom. On the other end was attached a straight-through high-vacuum in-line valve (Kontes 826600-0004). To seal the flask, the top consisted of a 2.5-cm o.d. tubing attached to a right-angle high-vacuum valve (Kontes 826610-0004) adapted to similar 1.3-cm o.d. tubing for connection to the trapping system. The top was connected to the incubation flask with a Cajon^R Ultra-Torr^R union (Cajon Company, Macedonia, OH, SS-16-UT-6). This union sealed the top to the flask body by compressing a Swagelok assembly and Viton O-ring against the glass tubing. Similar union fittings were used to connect the top of the flask to the trapping system.

Treatment of Soil Samples. Soil (30 or 50 g oven dry equivalent) was transferred into incubation flasks and moisture content adjusted to 1 bar with distilled water. The top assembly was then attached to the flask and the incubation flask was sealed shut. Samples were allowed to thermally equilibrate overnight in a darkened incubator set at 25 °C.

Samples were dosed with a solution of uniformly labeled ^{14}C-1,3-D which was delivered directly into the soil using a standard 100-μL syringe inserted through the Reacti-Vial closure. The application rate was 100 μg/g (equivalent to 388 kg/ha use rate for field crops). Following application, the samples were gently swirled by hand agitation to mix soil slightly and placed in a darkened incubator at 25 °C. Day 0 samples were sacrificed immediately and analyzed as soon as possible following dosing. The remaining samples were taken at pre-selected time points and were analyzed within 48 hours of sacrifice. Multiple trials were conducted for Catlin silt loam (3) and Fuquay loamy sand (2) to verify the ability of the incubation flask to maintain material balance. Only single kinetics trials were conducted for Wahiawa silty clay A-Horizon and B-Horizon.

Trap Assembly for Volatiles. The volatiles trap assembly used during sample workup consisted of a solid phase for adsorption of volatile organic compounds and a scrubbing tower liquid phase to trap CO_2 (Figure 1). Solid phase traps consisted of two sections of acid-washed, activated carbon separated by a small plug of glass wool in a heavy-walled glass tubing with several indentations about 1.5 cm from one end of the tube to provide support for the solid phase material. It is important to acid wash the activated carbon with 0.1 M HCl and air-dry the carbon over night so that CO_2 is not retained on it. The solid phase traps were connected to a heavy-walled glass tubing which had a male 18/9 spherical joint which could be connected to a scrubbing tower with a 18/9 female spherical joint. The tower was used to hold 0.2 M NaOH

Table I. Selected Physiochemical Properties of Soils

	Catlin Silt Loam	Fuquay Loamy Sand	Wahiawa Silty Clay, A	Wahiawa Silty Clay, B
pH	6.6	4.7	4.7	4.1
OC, %	2.0	0.6	2.3	0.8
CEC, (meq/100g)	16	3	10	6
1-bar moisture, (%)	20.0	3.8	26.5	32.2

Catlin silt loam: Fine-silty, Mixed, Mesic, Typic Argiudolls
Fuquay loamy sand: Loamy, Siliceous, Thermic, Arenic Plinthic Kandiudults
Wahiawa Silty Clay: Clayey, Kaolinitic, Isohyperthermic, Tropeptic Eutrustox
Wahiawa Silty Clay, A: 0 to 15 cm
Wahiawa Silty Clay, B: 60 to 75 cm

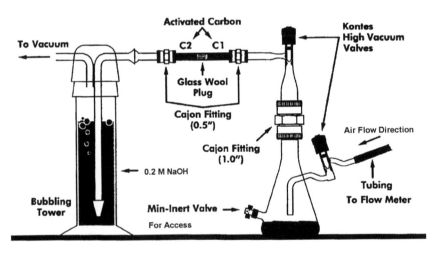

Figure 1.   Schematic for the Incubation Flask and the Headspace Purge of the Incubation Flask. The Incubation Flask is not Attached During Incubation.

solution to trap $^{14}CO_2$. The opposite side of the gas scrubbing tower was then connected to vacuum for purging.

Sample Sacrifice - Purge and Extraction. The headspace of incubation flasks was purged to collect volatiles before the flask was opened for the addition of extraction solvent. The flask top was connected to the volatile trap assembly described earlier. Vacuum was applied to the system to pull air through the flask. Before opening the lower Teflon valve, the upper valve was opened to form a partial vacuum in the flask thus preventing any volatiles from escaping out of the flask. The lower valve was then opened and the flask was purged with air for 30 minutes at 50 mL/minute (~12 times the flask headspace). The duration of the headspace purge was investigated and 30 minutes was shown to provide optimum recoveries of the applied 1,3-D.

After purging, acetone was aspirated into the soil flask through the lower flask inlet. The flask was resealed and agitated on a horizontal shaker for 2 hours. The soil and extract were then transferred to a centrifuge bottle and centrifuged for 20 minutes at 4000 rpm. The extract was decanted into a volumetric flask. A second acetone extraction was conducted and the combined extracts were brought to volume and assayed by LSC. Due to low ^{14}C activity, acetone soil extracts were concentrated at reflux in a round-bottomed flask equipped with a Snyder column. Recoveries of ^{14}C activity averaged greater than 98% for acetone extracts. The soil samples were then extracted with 0.2 M NaOH, centrifuged, and the extract brought to volume for LSC assay. After extraction with 0.2 M NaOH solution, soil samples were mixed, portions were weighed out and frozen in preparation for combustion. The frozen soil samples were combusted with an OX-300 R. J. Harvey combustion unit.

Immediately following purging, the carbon traps were extracted separately in centrifuge tubes with acetone. Samples were shaken on a horizontal shaker for 1 hr and the extract was decanted into a volumetric flask. A second extraction was conducted using the same conditions. The extracts from each trap were brought to volume and assayed by LSC. Less than 1% of applied 1,3-D was usually found in the second carbon trap.

Soil Respiration. Control samples of Fuquay loamy sand were incubated in the soil flasks to determine the rate of soil microbial respiration. Soil (30 g) was added to a flask and adjusted to one bar moisture and 150 mL of 0.2 N NaOH solution was added to a second flask. The flasks were connected via tubing so that they shared a common headspace and then incubated in the dark at 25 °C. Aliquots of the NaOH solution were analyzed at Days 0, 3, 7, 14, 35 and 70. The NaOH trapping solution (10 mL) was treated with solid barium chloride to precipitate carbonate and was then back-titrated with 0.1 M HCl solution (phenolphthalene as indicator).

Analysis of NaOH Extracts. Size exclusion chromatography was conducted on NaOH extracts to qualitatively determine molecular weight distribution of ^{14}C-labeled substances from acetone-extracted Catlin silt loam soil samples. Columns were prepared with Sephadex G-25 slurried in water and then poured into an open column (bed 2.5 cm i.d. x 31.5 cm long). One mL fractions of distilled water eluent were collected and mixed with Ultima Gold cocktail prior to LSC assay. The retention volumes of the materials in these extracts were compared to those of 3-chloroacrylic acid and 3-chloroallyl alcohol.

Metabolite Identification. Initial efforts to identify metabolites were conducted on acetone extracts from Catlin silt loam. Metabolites were investigated by HPLC and GC/MS techniques. Reverse phase HPLC was conducted on a Waters NovaPak C_{18} column to determine peak retention times for standards and metabolites. GC/MS analyses were conducted on underivatized concentrates.

Acetone extracts of Catlin silt loam were concentrated by either being blown down by a stream of nitrogen or with a Savant Automatic SpeedVac Concentrator. Each concentrate was analyzed by reverse phase HPLC (Gradient 1) on a NovaPak C_{18} column and the resulting retention times were compared to those generated from ^{14}C labeled standards.

Prior to derivatization, acetone concentrates from Catlin silt loam were analyzed by GC/MS on a Finnigan Model 9611 capillary chromatograph coupled to a Finnigan Model 4615 mass spectrometer with a Superincos data system. GC conditions were: hold for 2 minutes at an initial temperature of 50 °C, ramp to 250 °C at a rate of 10 °C/minute, and hold at 250 °C for 5 minutes. The column was a J & W Scientific DBFFAP, 30 M x 0.25 mm x 0.25 mm film thickness.

Additional soil samples of Catlin silt loam and Fuquay loamy sand were treated with ^{14}C-radiolabeled 1,3-D (specific activity 1.15 mCi/mmole) to facilitate the tracking of minor metabolites by HPLC analyses. In addition, control samples were treated with acetone only. Soil moisture was adjusted to 1 bar with distilled water. The 1,3-D treated and control samples were incubated in the dark at 25 °C for 14 days for the Catlin silt loam and 49 days for the Fuquay loamy sand (about one half-life for each soil).

The 1,3-D treated soil samples were purged as before and the carbon traps were extracted with acetone as before. Subsamples from the two soils were then extracted with acidified acetone solution (60% acetone and 40% 1 M HCl by volume) to enhance extraction from the soil organic matrix and then extracted twice with a 0.2 M NaOH solution. Control samples for Catlin silt loam and Fuquay loamy sand were extracted with acidified acetone and then 0.2 M NaOH solution. After neutralizing the HCl with NaOH solution, the extract was concentrated at reflux in a round bottomed flask equipped with a Snyder column. The resulting aqueous concentrate was concentrated to 2.0 mL under a stream of nitrogen gas. The resulting solutions were derivatized with acidic 1-butanol and the remaining water was azeotroped from the solution at the time of derivatization.

Derivatization of Metabolites. Derivatization of metabolites and carboxylic acid standards was accomplished by esterification with 1-butanol with sulfuric acid (3.8% by volume) as catalyst. Typically, 20 to 40 mg of the carboxylic acid was mixed with 1 mL of the acidified 1-butanol solution and then heated on a Pierce Reacti-Therm heating module for six hours at 105 °C. The reaction mixture was allowed to cool, mixed with 1 mL of hexane and extracted at least three times with a $KHCO_3$ solution, then washed several times with distilled water, and the hexane solution was dried over anhydrous sodium sulfate. ^{14}C-labeled butyl ester standards were prepared from unlabeled carboxylic acids and ^{14}C-labeled 1-butanol. In addition, several ^{14}C-labeled carboxylic acids were reacted with acidified 1-butanol solution which allowed estimates of reaction efficiencies to be obtained. The ^{14}C-labeled esters were used as standards for HPLC and GC/MS retention times and mass spectral fragmentation patterns.

Extract solutions containing metabolites were concentrated prior to derivatization with acidified 1-butanol. Acetone extracts were concentrated with a Snyder column. The remaining solution was adjusted to pH 7 with base. The resulting solution was then either blown down under nitrogen or concentrated by azeotroping the water with benzene. When benzene was used, the benzene solution was concentrated to a minimal volume and then mixed with acidified 1-butanol solution. The resulting mixture was heated at 105 °C for 6 hours. The resulting derivatized mixtures were mixed with hexane and extracted with $KHCO_3$ solution three times and then water three times. The hexane solution was dried over sodium sulfate and then concentrated under nitrogen for analysis by either HPLC or GC/MS.

GC/MS Analyses. A Hewlett Packard GC/MS was used to analyze the butyl esters of carboxylic acid standards and derivatized metabolites. A Hewlett Packard computer (9561A) equipped with HP59970C GC/MS Workstation revision 3.2 was used to interface with the GC Model 5890A Series II and MSD Model 5971A. The GC conditions were held at 50 °C for 5 minutes, ramped (20 °C/minute) to a final temperature of 280 °C and held for 5 minutes. A 15 M x 0.25 mM x 0.25 m J & W Scientific DB-5 GC column was used to conduct the analyses. The injection port and detector temperatures were 200 °C and 280 °C, respectively.

Results and Discussion.

Distribution of ^{14}C Activity. The initial ^{14}C activity was partitioned into multiple compartments. A portion of the ^{14}C activity was readily volatilized from the soils in the incubation flasks and was trapped on activated carbon or in 0.2 M NaOH trapping solutions. Another portion of ^{14}C activity was extracted from the soil samples by acetone extraction. A third fraction was extracted by 0.2 M NaOH solution, and the final compartment contained unextractable residues which were quantitated by combustion of the extracted soils. Partitioning of ^{14}C activity for Catlin silt loam and Fuquay loamy sand are shown in Tables II and III. A continual decrease in ^{14}C activity in the carbon traps was associated with a decrease of volatile 1,3-D and an increase of CO_2 in the caustic trapping solutions. Acetone extractable, caustic extractable, and unextractable ^{14}C activity tended to increase throughout the incubations as degradates were formed. For Catlin silt loam, the acetone extractable ^{14}C activity reached a maximum at Day 7 and then decreased. Similar partitioning of ^{14}C activity for Wahiawa silty clay A- and B-Horizons was reported previously (7).

Mass Balance. These laboratory studies demonstrated the ability of the incubation system to maintain material balance for a volatile compound such as 1,3-D and to trap 1,3-D and CO_2 from a closed system. Mass balance was determined from the sum of the ^{14}C activity recovered from the purge of the incubation flask's headspace, extraction of the soils and combustion of the extracted soils. From these results, the sampling procedures were shown to be acceptable because the mass balance for all experiments averaged from 81 to 94% although studies of longer duration tended to have lower average mass balance and greater standard deviations (Table IV).

All experiments showed a gradual loss of ^{14}C activity at later times. For the Catlin silt loam, the only low recovery was obtained for one sample which came after more than two half lives of 1,3-D. On the Fuquay loamy sand, recoveries to DAT 70 were acceptable considering the high volatility of 1,3-D. The average recovery through 42 days (close to one half life) was 86% ± 10%. Similar losses of ^{14}C activity were observed with Wahiawa silty clay A- and B-Horizon. Some of the observed decline may have resulted from O-ring failure which might have allowed volatile compounds such as 1,3-D and/or CO_2 to escape. Even with the slow decline of total ^{14}C activity, the decline of 1,3-D was clearly defined for all soils and these samples gave reliable data for the determination of a 1,3-D half-life.

Degradation Kinetics. The degradation of 1,3-D in aerobic soils was modeled as a first order decay process. The amount of 1,3-D remaining at each sacrifice time was determined from the sum of 1,3-D found from the flask purge and in the acetone soil extract. The data on the half-life of 1,3-D are summarized in Table V.

In the case of Wahiawa silty clay A-Horizon (7), 1,3-D degradation did not correspond well to a first order decay process ($r^2 = 0.77$) because of the rapid decay of 1,3-D, but appeared to fit better to a "Monod-with-growth model" ($r^2 = 0.88$). The rapid soil half-life (1.8 days) for Wahiawa silty clay A-Horizon may be an example of soil in which biotic degradation of 1,3-D occurs along with abiotic hydrolysis as the

Table II.   Catlin Silt Loam: Averaged Distributions and Recoveries as Percent of Applied

Day	Carbon Trap	Caustic Trap	Acetone Extract	Caustic Extract	Soil Combustion	Overall Recovery
0	95.0	0.3	2.5	0.8	0.6	99.0
1	75.4	1.5	6.9	2.8	3.9	90.4
3	65.8	3.0	8.5	4.7	8.0	89.9
7	52.1	5.1	12.0	8.2	13.4	90.6
11	38.8	9.1	10.6	9.4	16.5	84.3
15	33.2	14.0	9.6	11.5	22.1	90.3
20	23.1	14.7	9.5	13.3	27.6	88.1
26	15.2	20.4	9.0	14.0	30.4	88.8
30	10.8	19.4	5.8	13.9	27.6	77.4
					Average:	88.8
					St. Dev.	5.7

Table III. Fuquay Loamy Sand: Averaged Distribution and Recoveries as Percent of Applied ^{14}C

Day	Carbon Trap	Caustic Trap	Organic Extract	Caustic Extract	Soil Combustion	Overall Recovery
0	96.1	0.1	1.9	0.2	0.1	98.3
3	83.3	0.3	9.7	1.7	1.6	96.5
7	76.3	0.4	10.6	3.6	4.1	94.8
14	66.2	0.4	12.1	4.9	4.5	88.1
28	54.9	0.6	13.8	7.4	7.1	83.7
42	46.4	1.1	15.8	7.7	8.1	79.0
70	37.6	1.6	18.2	19.4	17.7	94.4
105	25.9	2.1	19.3	13.9	10.6	71.6
					Average	88.3
					St Dev	9.5

first step in the degradation of 1,3-D. The first order degradation of 1,3-D on Wahiawa silty clay B-Horizon yielded a 18.5 day half-life ($r^2 = 0.98$). For the Catlin silt loam, 1,3-D half-lives of 12.7 and 12.3 days were observed for Test 2 and Test 3 which were comparable to rates reported previously for loam soils (3, 9). The decline of 1,3-D on Catlin silt loam is shown in Figure 2. Fuquay loamy sand Tests 1 and 2 yielded 1,3-D half-lives of 35 and 61 days, respectively (Figure 3). Several factors may have contributed to the longer half-lives for this soil. This soil had been stored for nine months prior to initiation of the second trial which may have reduced its viability. Another possible explanation for the longer half-life on the Fuquay loamy sand may be that degradation was hindered by the high initial concentration of 1,3-D (~100 µg/g). The product label indicates that mineral soils should be treated at no more than 195 kg/ha which corresponds to a laboratory application rate of 50 µg/g. Inhibition of 1,3-D degradation has been observed in laboratory studies conducted at 62 µg/g but not at 12 µg/g (8). Smelt, Teunissen, Crum, and Leistra (5) have also reported suppression of 1,3-D degradation rates on loamy soils at concentrations of 470 µg/g as compared to rates at 3.7, 18, and 98 µg/g. Given the difficulties in accurate determination of soil moisture release curves for soils of high sand content, the soil moisture for 1 bar may have been imprecise and led to the use of a lower soil moisture which could have affected the rate of degradation.

Identification of Metabolites. Soil extracts were extensively investigated to characterize the aerobic soil metabolites of 1,3-D. HPLC methods included reverse phase and ion exclusion chromatography. Derivatization was conducted in preparation for GC/MS and reverse phase HPLC analyses of the derivatives.

Identification of Major Metabolites. Analysis of 1,3-D metabolites by reverse phase HPLC on a C_{18} column resolved 3-chloroallyl alcohol, *cis*-3-chloroacrylic acid and *trans*-3-chloroacrylic acid (Figure 4). The peak observed at 6 to 8 minutes was shown to contain additional polar metabolites. Ion exclusion HPLC provided further confirmation of 3-chloroallyl alcohol, *cis*-3-chloroacrylic acid and *trans*-3-chloroacrylic acid (Figure 5).

GC/MS analysis confirmed that 1,3-D, 3-chloroallyl alcohol, and 3-chloroacrylic acid were present in extracts of Catlin silt loam and Fuquay loamy sand. For 1,3-D, mass spectra were obtained with the expected ion ratios at $m/z = 110, 112,$ and 114. The mass spectra of 3-chloroallyl alcohol (standard and metabolite) exhibited the expected chlorine isotopic ratio for a single chlorine at $m/z = 75/77$ and $91/93$. The mass spectra of 3-chloroacrylic acid (standard and metabolite) were comparable with molecular ions observed at $m/z = 106/108$.

The butyl esters of 3-chloroacrylic acid (standard and metabolite) were observed in derivatized acetone extracts from both soils with GC/MS retention times at 7.92 and 7.89 minutes for butyl *cis*-3-chloroacrylate standard and derivatized metabolite, respectively. The GC/MS retention times for the butyl *trans*-3-chloroacrylate standard and derivatized metabolite were 7.21 and 7.15 minutes, respectively. The fragmentation patterns for the butyl esters of the standards and the metabolites compared favorably with ion fragments at $m/z = 89/92$ and $107/109$.

Identification of Minor Metabolites. Ion exclusion HPLC, derivatization, and GC/MS analyses resulted in the identification of additional metabolites from the Catlin silt loam and Fuquay loamy sand. From ion exclusion HPLC, a total of at least 11 to 12 distinct metabolites (from ~0.1 to 0.4% of applied) were observed from extracts of the Catlin silt loam (Figure 5) and the Fuquay loamy sand. Confirmation of acetic acid, propionic acid, and oxalic acid were obtained from the ion exclusion HPLC of Catlin silt loam and Fuquay loamy sand extracts.

Reverse phase HPLC was conducted on the butyl esters of metabolites (Figure 6) and ^{14}C-butyl ester standards. HPLC retention times for the butyl esters

Table IV. Average Recovery of ^{14}C Activity

Soil Study	Study Duration (days)	Average Recovery (%)	Standard Deviation (%)
Catlin Test 1	7	94.2	8.0
Catlin Test 2	14	90.2	2.4
Catlin Test 3	30	88.8	5.7
Fuquay Test 1	14	88.7	3.9
Fuquay Test 2	105	88.3	9.5
Wahiawa A-Horizon	44	81.0	17.8
Wahiawa B-Horizon	44	84.6	16.7

Table V.   Summary of 1,3-D Half Life on Tested Aerobic Soils

Soil	Half Life (days)	R^2
Catlin Silt Loam Test 2	12.7	1.00
Catlin Silt Loam Test 3	12.3	0.99
Fuquay Loamy Sand Test 1	35	0.93
Fuquay Loamy Sand Test 2	61	0.98
Wahiawa Silty Clay, A-Horizon	1.8	0.88
Wahiawa Silty Clay, B-Horizon	18.5	0.98

Degradation of 1,3-D: Catlin Silt Loam

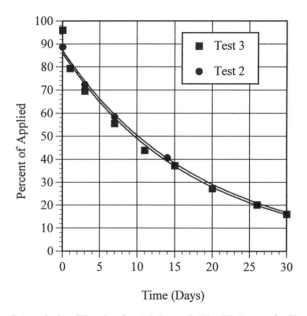

Time (Days)

Figure 2.   Degradation Kinetics for 1,3-D on Catlin Silt Loam for Tests 2 and 3

Degradation of 1,3-D: Fuquay Loamy Sand

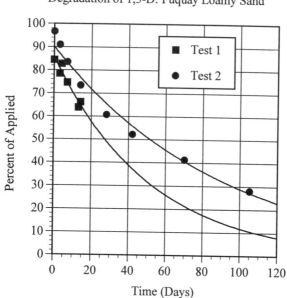

Figure 3.   Degradation Kinetics for 1,3-D on Fuquay Loamy Sand for Tests 1 and 2

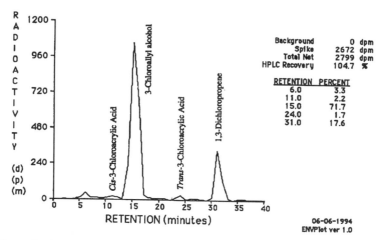

Figure 4.   Reverse Phase HPLC of an Acetone Extract; Day 105 from Fuquay Loamy Sand

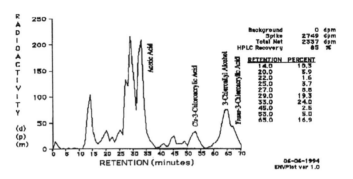

Figure 5.   Ion Exclusion HPLC of Carboxylic Acid Metabolites from Catlin Silt Loam

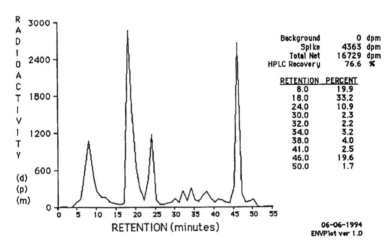

Figure 6.   Reverse Phase HPLC of Carboxylic Acid Metabolites as Butyl Esters - Day 14 from Catlin Silt Loam

are presented in Table VI. From these results, it is apparent that several standards co-elute. Butyl esters of the minor metabolites also co-eluted with possible standards. Investigation of soil extracts by GC/MS was conducted on underivatized and derivatized samples from 1,3-D treated and control samples of the Catlin silt loam and Fuquay loamy sand. Acetic acid and propionic acid were detected in Catlin silt loam extracts by GC/MS analyses. The mass spectra of propionic acid had fragments at $m/z = 45$, 57, and 74. The GC/MS data obtained for the butyl ester derivatives are summarized in Table VII.

The carboxylic acids identified in soil extracts from Catlin silt loam and Fuquay loamy sand can be rationalized as the result of 1,3-D treatment. The proposed scheme for the degradation of 1,3-D is shown in Figure 7. The initially formed 3-chloroallyl alcohol is oxidized either chemically or biochemically to 3-chloroacrylic acid which is then degraded biochemically (19, 20). Soil cultures of *Pseudomonas* degraded 3-chloroallyl alcohol and 3-chloroacrylic acid to 3-oxopropionic acid which could be oxidized to malonic acid or decomposed to acetaldehyde (19). Once present, acetaldehyde might be oxidized to acetic acid which is utilized in pathways such as the tricarboxylic acid (Krebs) cycle or the fatty acid synthesis. Therefore, most of the minor metabolites identified can be thought of as secondary metabolites of 3-chloroallyl alcohol and/or 3-chloroacrylic acid, the primary metabolites of 1,3-D.

Metabolite Distribution. As shown in Tables II and III, the distribution of ^{14}C activity in various compartments changed over the course of the 1,3-D incubations. The identitiy of the ^{14}C activity in the carbon traps was determined to be 1,3-D which decreased throughout the incubation of 1,3-D and [14]CO_2 concentration was found to increase in the caustic trapping solutions. The ^{14}C activity in acetone and NaOH extracts of soil increased with time and contained 3-chloroallyl alcohol, 3-chloroacrylic acid, and several other carboxylic acids. A slow rise of 3-chloroallyl alcohol was observed for the Catlin silt loam which attained 5% of applied by the end of the study (Table VIII). For the Fuquay loamy sand, the degradation of 3-chloroallyl alcohol was much slower and its soil concentration did not decline but attained 22% of applied at Day 105 (Table VIII). This result is similar to that observed by Roberts and Stoyden (6) on a loamy sand where 3-chloroallyl alcohol attained > 20% of applied after one month. From other work, the rate of degradation of 3-chloroallyl alcohol is often rapid on topsoils (< 2 to 4 days) (9). The biological degradation of 3-chloroallyl alcohol may have been affected by the reduction of O_2 levels in the sealed test system. Both this work and that of Roberts and Stoyden (6) were conducted in sealed systems where O_2 levels were not replenished. The soil microbial respiration rate was observed to level off about 42 days after treatment on Fuquay loamy sand. Therefore, results after about 42 to 70 days may not be representative of what the actual fate of 1,3-D and its degradates would be in the environment.

The degradation of the *cis*- and *trans*-3-chloroacrylic acid on Catlin silt loam occurred at different rates with the *cis*-isomer degrading more rapidly than the *trans*-isomer. This was inferred from the concentrations of the *cis*-isomer (maximum attained, 4% of applied) which peaked earlier than the *trans*-isomer (maximum attained, 6% of applied) at 7 and 20 days, respectively. Other authors have also observed this trend (3, 9).

Carbon dioxide was observed as the end product in the mineralization of 1,3-D. The Catlin silt loam produced carbon dioxide with a maximum of 22% of applied (Table VIII). On the other hand, the Fuquay loamy sand was less active and it only produced less carbon dioxide (max. of 2% of applied) (Table VIII). For Wahiawa silty clay, the A-Horizon and B-Horizon produced 37% and 1% of applied ^{14}C as [14]CO_2, respectively (Table VIII).

Table VI.   HPLC Data on Butyl Ester Derivatives

Butyl Ester	Soil	Present in Control Soils?	Standard HPLC RT (Minutes)	Metabolite Butyl Ester at Same RT?[c]
Acetate	Both[a]	Yes	13	Yes
Adipate	Both	Yes	39	Yes
Butyrate	Both	Yes	32	Yes
Chloroacetate	Catlin silt loam	No	12 to 14	Yes
4-Chlorobutyrate	Catlin silt loam	No	30	Yes
Fumarate	Both	Yes	41	Yes
Glycolate	Catlin silt loam	No	NA	Unknown
Hexanoate	Both	Yes	38	Yes
Lactate	Catlin silt loam	No	NA	Unknown
Malate	Catlin silt loam	No	20	Yes
Malonate	Both	Yes	28	Yes
2-Methylmalonate	Catlin silt loam	No	38	Yes
Oxalate	Both	Yes	29 to 30	Yes
Propionate	Catlin silt loam		19	Yes
Succinate	Both	Yes	30	Yes
Unk.[b]	Catlin silt loam	No	NA	Unknown

[a] Catlin silt loam and Fuquay loamy sand
[b] Chlorinated Unknown
[c] A positive response indicates that there is a butyl ester of a metabolite present at the same retention time as one of the standards.

Table VII. Mass Fragmentation Patterns of Butyl Ester Derivatives

Butyl Ester	Major Ion Fragments [c]
Butyl Acetate	<u>43</u>[c], 56, 61, and 73
Butyl Adipate	55, 87, 100, 101, 111, <u>129</u>, 143, 156, 185
Butyl Butyrate	56, <u>71</u>, and 89
Butyl Chloroacetate[a]	56, 57, <u>77</u>, 79, 95, and 97
Butyl 4-Chlorobutyrate[a]	56, 77, 79, <u>105</u>, 107, 123, and 125
Butyl Fumarate	56, 82, 99, <u>117</u>, 155, and 173
Butyl Glycolate	56, <u>57</u>, 61, and 77
Butyl Hexanoate	56, 99, and <u>117</u>
Butyl Lactate	<u>45</u>, 57, 75, and 85
Butyl Malate	57, 71, <u>89</u>, 117, 145, and 173
Butyl Malonate	57, 87, <u>105</u>, 143, and 161
Butyl 2-Methylmalonate	57, 74, 101, <u>119</u>, 157, and 175
Butyl Oxalate	<u>57</u>, weak ions at 101, 103, and 147
Butyl Propionate	56, <u>57</u>, and 75 weak ion at 87
Butyl Succinate	56, 57, <u>101</u>, 119, 157, and 175
Unk.[b]	57, 75, 101, <u>137</u>, 139, 193, and 195

[a] Exhibited chlorine isotopic pattern
[b] Chlorinated Unknown
[c] Base ion underlined

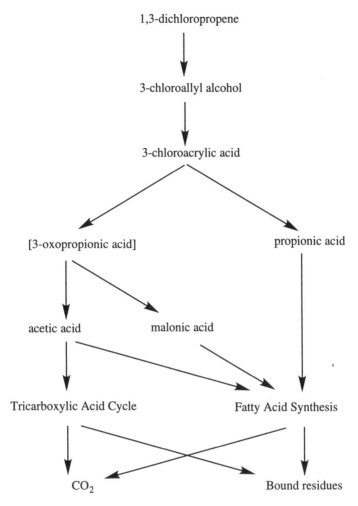

Figure 7. Proposed Scheme for the Degradation of 1,3-D on Aerobic Soil

Table VIII. Average Maximum Percent of Applied Attained by 1,3-D Degradates

	Catlin Silt Loam	Fuquay Loamy Sand	Wahiawa Silty Clay, A-Horizon	Wahiawa Silty Clay, B-Horizon
3-Chloroallyl Alcohol	5	23	≤1.5	26
cis--3-Chloroacrylic Acid	4	1	≤1.5	NA
trans--3-Chloroacrylic Acid	6	1	≤1.5	NA
Other Carboxylic Acids	4	5	NA	NA
Carbon Dioxide	22	2	37	1
NaOH Extractable	14	20	10	12
Unextractable	30	18	38	8

CONCLUSIONS

The degradation rate of 1,3-D was determined for Catlin silt loam, Fuquay loamy sand, and Wahiawa silty clay (A- and B-Horizon) in sealed incubation flasks. In a Catlin silt loam, 1,3-D dissipated rapidly with a half-life of 12 days. In a Fuquay loamy sand, 1,3-D had a half-life of 61 days. The 1,3-D half-life in Wahiawa A-Horizon was 1.8 days and in Wahiawa B-Horizon it was 18.5 days (7). The results for the Wahiawa A- and B-Horizon soils clearly show that there are differences in the rates of degradation of 1,3-D in top soil and subsurface soil and that degradation of 1,3-D can be significant in subsurface soil as well (18.5 days).

The observed degradation rates and metabolite concentrations (maximums) are artificial in that dissipation by volatilization was not taken into account in the experimental design. Field studies have shown that up to approximately 25% of the applied 1,3-D had volatilized after 14 days (16). In addition, the laboratory application rates for 1,3-D that were used in this study were equivalent to twice the maximum rates for field crops on mineral soils (1).

Soils treated with 1,3-D contained residues which partitioned into multiple compartments: easily volatilized residues, easily extracted residues with acetone, more difficultly extracted residues with 0.2 N NaOH, and unextractable residues. ^{14}C labeled materials were easily volatilized from all test soils and consisted of 1,3-D and CO_2. Complete mineralization to carbon dioxide was observed with 22% of applied ^{14}C activity for the Catlin silt loam, 2% for the Fuquay loamy sand, 37% for Wahiawa silty clay A-Horizon and 1% for Wahiawa silty clay B-Horizon. A portion of ^{14}C labeled residues was extractable with organic extractants and consisted of 1,3-D, 3-chloroallyl alcohol, cis-/trans-3-chloroacrylic acid, and numerous carboxylic acids (all < 0.4% of applied).

For Catlin silt loam and Fuquay loamy sand, numerous carboxylic acid metabolites were observed and were at levels less than 0.4% of applied. These carboxylic acids were identified as acetic acid (both), adipic acid (both), butyric acid (both), chloroacetic acid (Catlin only), 4-chlorobutyric acid (Catlin only), fumaric acid (both), glycolic acid (Catlin only), hexanoic acid (both), lactic acid (Catlin only), malonic acid (both), 2-methylmalonic acid (Catlin only), oxalic acid (both), propionic acid (both), and succinic acid (both). There were greater amounts of acetone extractable carboxylic acids at the later sampling times for both soils possibly due to reduced O_2 levels in the sealed incubation flasks. Additional ^{14}C labeled residues were extracted with NaOH solutions and were identified as 1,3-D, 3-chloroallyl alcohol, cis- and trans-3-chloroacrylic acid, and other carboxylic acids: 4-chlorobutyric acid (Catlin only), fumaric acid (both), malic acid (Catlin only), malonic acid (both), oxalic acid (both), and succinic acid (both). In addition to these substances, a portion of the ^{14}C label was associated with high molecular weight materials. Size exclusion chromatography showed that the NaOH extracts contained ^{14}C-labeled components that had low retention volumes and thus had higher molecular weights than compounds such as 3-chloroallyl alcohol (Figure 8).

The detection of 1,3-D in extracts (0.2 M NaOH) from soils previously extracted with acetone indicated that 1,3-D had become associated with the soil matrix. Furthermore, the amount of "bound" 1,3-D increased with continued incubation. The increased association of 1,3-D and other similar volatile halogenated hydrocarbons with the soil matrix has been documented for soils that have been aged with volatile organic compounds (14, 15, 23). Recent work of Pignatello (14, 15) showed that volatile halogenated hydrocarbons associate with the organic matter in the soil matrix. The outcome of this is that with prolonged contact with soil 1,3-D would be expected to be less mobile.

The remaining ^{14}C labeled material in the soils which was unextractable was quantified by combustion of the soil samples. For the Catlin silt loam, the unextractable material attained 31% of applied ^{14}C activity. The maximum

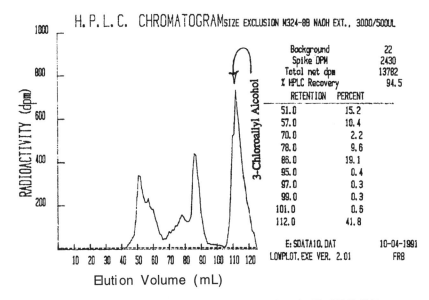

Elution Volume (mL)

H. P. L. C. CHROMATOGRAM SIZE EXCLUSION N324-8B NAOH EXT., 3000/500UL

Background	22
Spike DPM	2430
Total net dpm	13782
% HPLC Recovery	94.5

RETENTION	PERCENT
51.0	15.2
57.0	10.4
70.0	2.2
78.0	9.6
86.0	19.1
95.0	0.4
97.0	0.3
99.0	0.3
101.0	0.6
112.0	41.8

E: SOATA1O. DAT   10-04-1991
LOWPLOT. EXE VER. 2.01   FR8

Figure 8.   Size Exclusion Liquid Chromatography of a NaOH Soil Extract

unextractable ^{14}C labeled material on Fuquay loamy sand attained 18% of applied ^{14}C activity. Unextractable residues in Wahiawa silty clay A-Horizon and B-Horizon attained 38% and 8% of applied ^{14}C activity, respectively.

The observed metabolites of 1,3-D should not have a significant impact on the environment. The major metabolites, 3-chloroallyl alcohol and 3-chloroacrylic acid are known to rapidly degrade in aerobic soils (9, 10) and field data has shown that 3-chloroallyl alcohol levels (< 300 ppb) remained low in a field near the collecting site of the Fuquay loamy sand used to conduct the present aerobic soil metabolism study (24). Therefore, the higher levels of the 3-chloroallyl alcohol observed in this study (~23% of applied on Fuquay loamy sand) might be the result of the artificial nature of the sealed incubation vessels which were necessary to maintain mass balance. Most of the other minor metabolites were naturally occurring carboxylic acids which can be found in soil (25) and would be expected to undergo rapid metabolism in aerobic soil environments. The chlorinated acids, chloroacetic acid and 4-chlorobutyric acid, should also undergo rapid degradation.

References

1. DowElanco Product Label for Telone II, **1995**.
2. McCall, P. J. *Pestic. Sci.* **1987** *19*, 235-242.
3. Van Dijk, H., *Agro-Ecosystems* **1974** *1*, 193-204.
4. Van Dijk, H., *Pestic. Sci.* **1980** *11*, 625-632.
5. Smelt, J. H., Teunissen, W.; Crum, S. J. H., Leistra, M., *Netherlands Journal of Agricultural Science* **1989**, *37*, 173-183.
6. Roberts, T. R., Stoydin, G., *Pestic. Sci.* **1976**, 7, 325-335.
7. Wolt, J. D., Holbrook, D. L., Batzer, F. R., Balcer, J. L., Peterson, J. R., *Acta Horticulturae* **1993**, *334*, 361-371.
8. van der Pas, L. J. T.; and Leistra, M., *Arch. Environ. Contam. Toxicol.* **1987**, *16*, 417-422.
9. Leistra, M., Groen, A. E., Crum, S. J. H., van der Pas, L. J. T. *Pestic. Sci.* **1991**, *31*, 197-207.

10. Ou, L.-T., Chung, K.-Y., Thomas, J. E., Obreza, T. A., and Dickson, D. W., *J. Nematology* **1995**, *27*, 249-257.
11. Verhagen, C., Smit, E., Janssen, D. B., van Elsas, J. D., *Soil Biol. Biochem.* **1995**, *27*, 1547-1557.
12. McCall, P. J. **1986**. Unpublished results of DowElanco.
13. Tuazon, E. C.; Atkinson, R.; Winer, A. M., Pitts, J. N. Jr., *Arch. Environ. Contam. Toxicol.* **1984**, *13*, 691-700.
14. Pignatello, J. J., *Environ. Toxicol. Chem.* **1990**, *9*, 1107-1115.
15. Pignatello, J. J., *Environ. Toxicol. Chem.* **1990**, *9*, 1117-1126.
16. Knuteson, J. A., Petty, D. G., Shurdut, B. A. **1992**. Unpublished results of DowElanco.
17. Maddy, K. T., Fong, H. R., Lowe, J. A., Conrad, D. W., Fredrickson, A. S., *Bull. Environm. Contam. Toxicol.* **1982**, *29*, 354-359.
18. Obreza, T. A., Ontermaa, E. O., *Soil and Crop Sci. Soc. Fla. Proc.* **1991**, *50*, 94-98.
19. Besler, N. O.; and Castro, C. E., *J. Agric. Food Chem.* **1971**, *19*, 23-26.
20. Van Hylckama Vlieg, J. E. T., Janssen, D. B., *Biodegradation* **1992**, *2*, 139-150.
21. Karris, G. C., Downey, J. R. **1987a**. Unpublished results of The Dow Chemical Company.
22. Karris, G. C., Downey, J. R. **1987b**. Unpublished results of The Dow Chemical Company.
23. Steinberg, S. M., Pignatello, J. J., Sawhney, B. L., *Environ. Sci. Technol.* **1987**, *21 (12)*: 1201-1208.
24. Oliver, G. R.; Bjerke, E. L.; Woodburn, K. B.; and O'Melia, F. C. **1988**. Unpublished results of DowElanco.
25. Fox, T. R., Comerford, N. B, *Soil Sci. Soc. Am. J.* **1990**, *54*, 1139-1144.

Chapter 8

Environmental Fate of Chloropicrin

S. N. Wilhelm[1], K. Shepler[2], L. J. Lawrence[3], and H. Lee[4]

[1]Niklor Chemical Company, 2060 East 220th Street,
Long Beach, CA 90810
[2]PTRL West Inc., 4123B Lakeside Drive, Richmond, CA 94806
[3]PTRL East Inc., 3945 Simpson Lane, Richmond, KY 40475
[4]Bolsa Research Associates, 8770 Highway 25, Hollister, CA 95024

Chloropicrin environmental fate and residue studies carried out under EPA guidelines demonstrate that this preplant soil fumigant is readily metabolized in agricultural soil. The half-life of [^{14}C]chloropicrin was 4.5 days in sandy loam soil with carbon dioxide being the terminal breakdown product. In an anaerobic aquatic/soil system, [^{14}C]chloropicrin was dehalogenated to nitromethane with a half-life of 1.3 hours. Transient mono and dichloronitromethane intermediate degradates were identified. In a plant metabolism study utilizing soil treated with [^{14}C]chloropicrin, the radiolabeled carbon was incorporated into numerous natural plant biochemical compounds ostensively via the plant's single carbon pool. The photolytic half-life of chloropicrin in water was 31.1 hours with the photoproducts being carbon dioxide, bicarbonate, chloride, nitrate, and nitrite.

Chloropicrin, as a soil fumigant, is used for its broad biocidal and fungicidal properties primarily in high value terrestrial crops such as strawberries, peppers, onions, tobacco, flowers, tomatoes, and nursery crops. It is injected into the soil at least six inches deep, fourteen days or more before planting. Four environmental fate and residue studies conducted under the USEPA Pesticide Assessment Guidelines have elucidated the nature and extent of chloropicrin degradation products in aerobic soil, anaerobic soil/water, plants grown in treated soil, and in a photolyzed aqueous solution.

Aerobic Soil Metabolism

In a study (1) conducted at PTRL West Inc. (Richmond, CA) the half-life of chloropicrin in agricultural soil was determined as well as the nature and extent of the formation of degradation products. A sandy loam soil treated with [^{14}C]chloropicrin (Figure 1) at a rate equivalent to 500 lb/acre was incubated under aerobic conditions at 25^{0}C in the dark for 24 days.

0097–6156/96/0652–0079$15.00/0

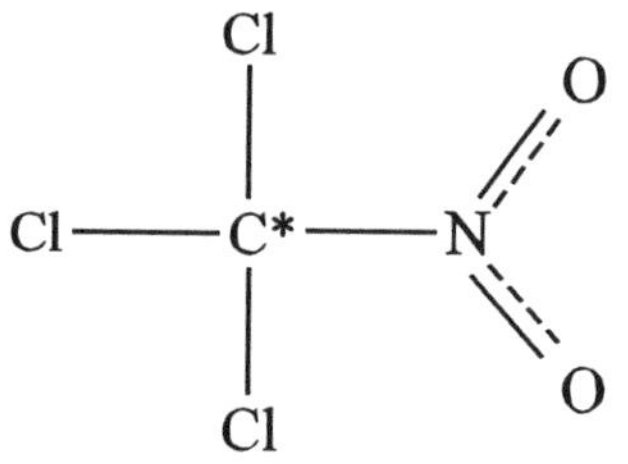

Cl
|
Cl————C*————N
| O
Cl   O

Molecular Weight 164.39
*position of ^{14}C label

Figure 1. Structure of [^{14}C]Chloropicrin

Test System. [^{14}C]Chloropicrin (98.3%) was obtained from New England Nuclear (Boston, MA) with a specific activity of 1.1 mCi/mmol. Soil was obtained from a strawberry field in Watsonville, California, and was classified as sandy loam according to the USDA texture classification system (Table I). Characterization of the test soil was performed by A & L Great Lakes Laboratories (Fort Wayne, IN).

Unlabeled reference chloropicrin (99.3%) was supplied by the Chloropicrin Manufacturers Task Force and nitromethane (>99%) was obtained from Aldrich Chemical Company. All solvents and reagents were reagent grade or better. Agars were obtained from DIFCO Labs (Midland, MI). All water used was HPLC grade or purified with a Barnstead Nano Pure II system (ASTM Type I).

Table I. Soil Physicochemical Characteristics

pH	7.2
CEC(meq/100g)	12.46
Organic content (%)	1.27
WHC % at 1/3 Bar	16.14
Sand (%)	53.2
Silt (%)	27.2
Clay (%)	19.6
USDA Texture Class	Sandy Loam
Bulk Density (g/cc)	1.12

The soil was passed through a 2 mm sieve to remove any debris present and soil moisture was adjusted with deionized water to 78% of field capacity at 1/3 bar prior to dosing. Fifty grams (dry weight) were added to each previously autoclaved 500 mL biometer flask. At the top of each flask was a horizontal sidearm containing a polyurethane foam (PUF) trap for organic volatiles. Connected to the sidearm was a liquid trap containing a thistle tube immersed in 40 mL of an aqueous 10% KOH solution. On the lower sidewall of each flask was a stopcock for dosing and removing of headspace samples. [^{14}C]Chloropicrin (100 μL) dissolved in acetonitrile was applied beneath the soil surface by inserting a gas-tight syringe through the sidewall stopcock fitted with a septum. Three series of flasks were dosed in this manner with concentrations of 271, 288, and 295 μg chloropicrin per gram of soil. The applied dose was taken to be the average concentration found upon immediate assay of the Time 0 samples.

The biometer flasks, wrapped in foil to exclude light, were incubated in water at 25^{0}C. To supply oxygen to the system, a regulated oxygen cylinder provided a slight positive pressure to the thistle tube in each KOH trap. To verify that aerobic conditions were maintained throughout the study, a surrogate flask containing a redox indicator solution of resazurin dye was connected to the flask manifold.

Sampling and Analysis. Sampling was performed after 4.5 hours and at 1, 2, 3, 6, 14, 21, and 24 days. At each sampling time, 2-4 flasks were analyzed. Aliquots of the headspace, soil, PUF and KOH traps were analyzed for total radiocarbon, chloropicrin and metabolites.

The flask headspace was sampled by taking three 5 mL aliquots through the stopcock/septum and injecting them into a Harvey OX-600 Biological Oxidizer for combustion to $^{14}CO_2$ and subsequent radioassayed by liquid scintillation counting (LSC). All LSC analyses were performed with a Beckman Liquid Scintillation Counter Model LS5000CE or LS6000C. One additional 5 mL headspace aliquot was injected into a septum fitted vial containing acetonitrile for subsequent analysis by HPLC and GC/MS.

To initiate soil extraction, chilled acetonitrile (~50 mL) was injected through the stopcock/septum. After thirty minutes the system was opened and the KOH and PUF traps removed. The KOH solution was analyzed by LSC before and after precipitation with $BaCl_2$ to quantitate $^{14}CO_2$. The PUF was extracted with acetonitrile and radioassayed by LSC in scintillation cocktail.

The soil was then transferred to pre-weighed 250 mL centrifuge bottles. Two additional aliquots of acetonitrile (50 mL each) were used to facilitate the transfer. The centrifuge bottles were shaken by hand for approximately two minutes and centrifuged. The supernatant was decanted and the soil extracted two additional times. The extracts were combined and aliquots were radioassayed by LSC. All HPLC analyses were performed within 24 hours. Following extraction, the soil pellet was weighed and either stored frozen or radioassayed immediately by combustion and LSC.

A humic acid/fulvic acid partition was performed on the post-extracted soil of the Day 14, 21 and 24 samples. A 5.0 g aliquot of the soil was placed in a shaker for 24 hours with 25 mL of 0.5 M NaOH. Following centrifugation, the supernatant was removed and the soil washed with an additional 12.5 mL of 0.5 M NaOH. Following another centrifugation the extracts were combined, washed, acidified to pH 1 with HCl and placed in an ice water bath to precipitate the humic acids. The extract was then centrifuged and the humic and fulvic acid fractions were separated. The humic fraction was redissolved in 0.5 M NaOH and both fractions were radioassayed by LSC.

HPLC analyses of the extracts were performed with a Perkin-Elmer Series 4 equipped with a LC90 UV Spectrophotometric Detector (220nm) or with a Perkin-Elmer Series 410 equipped with a LC-235 Diode Array Detector and Cygnet ISCO Fraction Collector. Both units used Beckman 171 Radioisotope Detectors and Beckman 110B solvent delivery systems. A Supelco LC-18 column was utilized with a linear gradient water/acetonitrile solvent system. HPLC recoveries were 98.2 $\pm$ 10.4%. GC/MS analyses were accomplished with a Hewlett Packard MS5971 in the electron impact mode. Separation was carried out with a Stabilwax column (crossbond Carbowax PEG, 30 m x 0.25 mm) at 35-220^{0}C.

The detection limit for soil samples upon combustion and LSC (2X background) was 0.067 ppm. For HPLC analyses with a sample size of 10,000 dpm in a matrix containing 0.2 ppm, the detection limit was 0.0012 ppm.

Kinetic Distribution of Radiocarbon in System Compartments. Radiocarbon was found quantitatively ($\geq$ 98%) in the soil at Time-0. Chloropicrin partitioned reversibly between the soil, headspace and foam plug during the study. The quantity of extractable radiocarbon in the soil decreased rapidly at first then more slowly to a total of 6.2% of applied after 24 days. Unextractable radiocarbon in the soil steadily increased to 14.7% of applied during the same period. Radiocarbon in the KOH trap increased from 1.6% after 4.5 hours to >70% of applied by Day 24. Overall recovery of radiocarbon for the study was 97.2 $\pm$ 6.0%.

Metabolic Half-life. The half-life of [^{14}C]chloropicrin in sandy loam soil at 25^{0}C was calculated to be 4.5 days based on pseudo-first order kinetics (r^2 = 0.946). The amount of chloropicrin present in the entire system (soil, headspace and traps) was included in the half-life expression since the compound partitioned reversibly between the compartments throughout the study.

Metabolites. After 24 days, 65.6 - 75.2% of the applied radiocarbon was present as [^{14}C]carbon dioxide. Two intermediates in the aerobic metabolism of chloropicrin to carbon dioxide were also identified. [^{14}C]Chloronitromethane was identified by GC/MS at up to 5.5% of the applied radiocarbon (Day 14) and nitromethane was identified by co-chromatography at levels up to 4.1% (Day 24). In a concurrent volatility study (2), dichloronitromethane, the first intermediate in the successive dehalogentation of chloropicrin to nitromethane, was also identified by GC/MS.

A metabolite eluting at the HPLC solvent front and consistent with the carbonate standard was present at up to 3.9% of applied by Day 24. Radiocarbon incorporation into the fulvic acid fraction increased to ~ 4% of applied by Day 24 while a negligible amount (<1%) was found in the humic acid fraction. Total unextractable soil radiocarbon increased steadily during the course of the study to 14.7% by Day 24.

The proposed metabolic pathway for the degradation of chloropicrin in aerobic soil is shown in Figure 2.

$$CCl_3NO_2 \longrightarrow HCCl_2NO_2 \longrightarrow H_2CCl\,NO_2 \longrightarrow$$

$$H_3CNO_2 \longrightarrow CO_2 \longrightarrow \begin{array}{c} \text{Incorporation into soil} \\ \text{and} \\ \text{Fulvic/Humic Acids} \end{array}$$

Figure 2. Proposed Metabolic Pathway of Chloropicrin in Aerobic Soil

Carbon dioxide, as the terminal breakdown product of chloropicrin, was also confirmed under field conditions in a concurrent volatility study (*3*). In the volatility study, four boxes (18in. x 14in. x 18in. deep) containing soil treated with [^{14}C]chloropicrin were fitted with frames covered with polyethylene film. Air was then drawn over the soil and through a KOH trap. The trapping solution was then radioassayed by LSC to determine [^{14}C]carbon dioxide. [^{14}C]Carbon dioxide emissions remained high for the first 14 days and then dropped off rapidly.

Soil Viability. The soil was analyzed for microbial viability prior to dosing, midway through the incubation period and after all sampling was complete. Aerobic bacteria were enumerated on trypticase soy agar, actinomycetes on actinomycetes isolation agar and fungi on potato dextrose agar.

Although low in organic content, the soil utilized in this study remained viable even after treatment with chloropicrin at a rate equivalent to 500 lb/acre. Since chloropicrin is used for its fungicidal properties, it was expected that fungi levels in the treated soil would be reduced significantly. Total fungi colony forming units/g (CFU/g) were reduced by three orders of magnitude from a pre-dosing average of 0.020 x 10^6 to 0.035 x 10^3 CFU/g by Day 29. Total aerobic bacteria levels dropped slightly by Day 12 but recovered to their initial levels (5 - 21.5 x10^6 CFU/g) by Day 26. Total actinomycetes were reduced by almost two orders of magnitude by Day 12 but recovered close to their initial level of 4.5 x 10^6 CFU/g by Day 29.

Anaerobic Aquatic/Soil Metabolism

In a study (*4*) conducted at PTRL West Inc., a mixture of soil flooded with water and dosed with [^{14}C]chloropicrin at a rate equivalent to 500 lb/acre was incubated at 25^0 C in the dark under anaerobic conditions. The soil, radiolabeled chloropicrin, and test apparatus utilized were the same as in the previously described study except that the system was rendered anaerobic prior to dosing and was supplied with nitrogen throughout the incubation period. The half-life of chloropicrin was determined as well as the nature and extent of the formation of degradation products.

Test System. Soil (20 g dry weight) and purified water (87 mL) were added to each previously autoclaved 500 mL biometer flask. Flasks were flushed periodically with nitrogen for 1-5 hours over several days to render the system anaerobic. To promote anaerobic conditions, 0.2 g of alfalfa were added to each flask. The flasks were maintained under anaerobic conditions four weeks prior to dosing the first set and for eleven weeks prior to dosing the supplemental set. During the course of the study, anaerobic conditions were monitored by measuring the redox potential (E_h) and oxygen content at the water/soil interface of randomly selected samples. In addition, to monitor the status of the test system, a surrogate flask containing test material, soil, water and resazurin redox indicator dye was maintained in line with the sample flasks. This flask was monitored throughout the study for the appearance of a pink color in the water which would indicate the presence of oxygen.

Since chloropicrin has a limited solubility in water (1.6 g/L)(*1*), special measures were required in order to deliver a quantifiable dose to the water/soil mixture. A radiolabeled stock solution was prepared by dissolving 100 µL of [^{14}C]chloropicrin in acetonitrile. The dosing solutions were then prepared by diluting 85 μL of the stock solution with 52 mL of deoxygenated water in each of six vials. The vials were equipped with Teflon lined septa and contained minimal headspace after filling. Acetonitrile, as a co-solvent, was present in the final dosing solutions at < 0.2%. The vials containing the dosing solutions were allowed to equilibrate overnight in a refrigerator.

To avoid losing chloropicrin to the headspace of the vials, 50 of the 52 mL in each vial were withdrawn at one time into a gas-tight syringe. Each biometer flask then received 10 mL of the dosing solution with an additional 10 mL being delivered to a surrogate dosing flask containing acetonitrile for verification of the nominal applied radiocarbon. Dosing of the biometer flasks was accomplished by inserting the syringe needle through the septa fitted stopcock and injecting the dose beneath the surface of the water/soil mixture. After application, the stopcock was closed, the septa removed and the flasks were swirled to ensure homogeneity of the mixture. Four flasks were treated from each vial of dosing solution. A total of 28 flasks were treated with an average concentration of 286 ppm. The application rate was determined by radioassaying aliquots of the surrogate vessels.

Sampling and Analysis. Samples were collected in duplicate immediately after dosing, after 1.5 hours and on Day 1, 2, 5, 12, 26, and 54. Aliquots of the headspace, water, soil, PUF and KOH traps were analyzed for total radiocarbon, chloropicrin and metabolites. Aliquots of the water phase were analyzed for the presence of oxygen.

Headspace aliquots (3 x 5 mL) were removed by gas-tight syringe via the septum/stopcock for injection directly into the oxygen inlet of the Harvey OX-600 Biological Oxidizer. The $^{14}CO_2$ generated was trapped with Carbon 14 Cocktail for subsequent radioassay by LSC. For the 1.5 hour sample, a fourth aliquot was injected into a septum fitted vial containing acetonitrile for HPLC analysis.

Aliquots of the water layer (3 x 1 mL) were also removed by syringe for radioassay. In addition, two aliquots of the water layer were removed and injected into septa fitted vials (previously flushed with nitrogen) for HPLC analysis and measurement of oxygen, pH, and redox potential (E_h). The flasks were then opened and the PUF and KOH traps were removed and radioassayed as described in the previous study.

The soil/water mixture was then transferred to 250 mL pre-weighed centrifuge bottles. The mixture was then centrifuged and the water phase decanted and stored in a freezer. The soil layer was extracted three times with acetonitrile and centrifuged. The extracts were combined and aliquots radioassayed by LSC. HPLC analyses of the extracts were usually performed on the same day as the extraction and always within 24 hours. The post-extraction soil was weighed and aliquots were assayed for radiocarbon. All samples and standards were stored under freezer conditions (< 0^0C) when not in use. Repeated injections of the reference standards indicated that no degradation occurred during the study.

The time required to perform Time 0 sampling of the headspace in the first set of flasks delayed analysis of the water/soil for approximately 90 minutes. For the second set of flasks, in order to acquire a sample that was closer to the actual Time 0, and since negligible quantities of radiocarbon were found during the initial headspace sampling, Time 0 sampling of the headspace was omitted.

HPLC analyses of the extracts were performed with a Perkin-Elmer Series 4 equipped with a LC90 UV Spectrophotometric Detector (220nm) and a Beckman 171

Radioisotope Detector (BFD); a Perkin-Elmer Series 410 equipped with a LC-235 Diode Array Detector, BFD and Cygnet ISCO Fraction Collector; or with a Hewlett Packard Series 1050 Iso Pump Model 020 equipped with a Variable Wavelength Detector and BFD or HP Series 1047 Refractive Index Detector.   Two HPLC columns were utilized: a Supelco LC-18 column with a linear gradient of acetonitrile in water and a Bio-Rad Amines Ion Exclusion HPX-87H column with 0.01N H_2SO_4 as solvent. HPLC recoveries were 86.1 $\pm$ 26.8% throughout the study.

GC/MS analyses were performed with a Hewlett Packard MS5971 in the electron impact mode.   Separation was carried out with a Stabilwax column (crossbond Carbowax PEG, 30 m length x 0.25 mm diameter) at 35-220^{0}C.

The detection limit for combusted soil samples (2X background) was 0.067 ppm. For HPLC radiochromatograms, the limit of detection was 0.0018 ppm.

Kinetic Distribution of Radiocarbon in System Compartments. Chloropicrin partitioned reversibly between the soil/water, headspace and foam plug during the study. Radiocarbon in the water layer decreased from ~90% of dose at Time 0 to ~20% at Day 54 .   Radiocarbon levels in the soil initially were low (~5%) and extractable levels decreased slowly over the incubation period to ~2% by Day 54.   However, the unextractable radiocarbon in the soil steadily increased to >30% of applied during the same period.   Radiocarbon in the KOH trap increased to ~50% of applied by Day 26. Only ~2% of the KOH radiocarbon, however, was confirmed by precipitation with $BaCl_2$ to be $^{14}CO_2$. The remainder was [^{14}C] nitromethane and its abiotic degradation products. The radiocarbon in the foam traps decreased from ~9% at 1.5 hours to <1% by Day 54. Significant levels of radiocarbon in the headspace were detected at the first two samplings (1.5, 4 hours) but decreased to <1% by Day 1.   Radiocarbon recoveries for the study were 98.2 $\pm$ 2.8%.

Metabolic Half-life. Chloropicrin degraded readily with a calculated pseudo first-order half-life of 1.3 hours ($r^2 = 0.999$).   Due to the rapid degradation, only three sampling times were utilized in the chloropicrin half-life calculation. The amount of chloropicrin present in the entire system (soil, water, headspace, traps) was included in the half-life expression since the compound partitioned reversibly between the system compartments throughout the study.

Metabolites. [^{14}C]Chloropicrin was metabolized to [^{14}C]nitromethane within hours. After one day, the average chloropicrin concentration was only 1.7% of applied. Over the same time period, [^{14}C]nitromethane increased to 58.7%.   These results are consistent with the published literature on the rapid biodehalogenation of chloropicrin by *Pseudomonads* (*5*).

[^{14}C]Chloronitromethane, an intermediate in the dehalogenation of chloropicrin, was present at 80.3% of the dose at the Time 0 sampling.   After four hours, the concentration of this intermediate was ~51%. [^{14}C]Carbon dioxide steadily increased to ~3.9% of dose by Day 54.   Bound radiocarbon in the post-extracted soil increased throughout the study to a maximum of 36.5% at Day 54.   The soil ^{14}C-fulvic acid fraction increased steadily to 24.4% of the dose by Day 54.   The ^{14}C-humic acid fraction contained $\leq$ 2% throughout the study.

[^{14}C]Chloropicrin was quantitated by LSC and confirmed by HPLC co-chromatography with a standard. Additionally, a hexane extraction of a water sample was analyzed by GC/MS revealing prominent ions at M + HCl (m/e 199), M + 1 (m/e 164) and M-NO$_2$ (m/e 117). Nitromethane was confirmed by HPLC co-chromatography using the C-18 reverse phase method. Secondary confirmation was obtained for the Day 2 sample using HPLC ion exclusion co-chromatography with a standard. Chloronitromethane was identified by GC/MS using the method for chloropicrin and revealed a band at 12.8 minutes with a base peak of m/e 49 (one chlorine pattern) corresponding to loss of the nitro group from chloronitromethane.

The proposed metabolic pathway for chloropicrin under anaerobic conditions is similar to what was proposed in the aerobic study (Figure 2.) with the first three steps being a successive dehalogenation to nitromethane. Nitromethane may then be incorporated directly into the soil and fulvic/humic acid fractions or it may be metabolized as carbon dioxide. As expected under anaerobic conditions, little of the radiocarbon (<4.4%) was found as carbon dioxide by Day 54.

Soil Viability and pH. The soil was analyzed for microbial viability prior to dosing, midway through the study and after all sampling was complete. Anaerobic bacteria were enumerated on anaerobic agar using a system obtained from DIFCO Laboratories (Midland, MI). The sandy loam soil utilized in this study remained viable throughout the incubation period. The pre-dosing anaerobic bacteria level of 11.95 x 10^6 CFU/g was reduced two orders of magnitude by the first analysis at Day 19 and remained unchanged throughout the remainder of the study.

Test system pH varied from a Time 0 average of 7.66 to a low of 6.37 four hours after dosing. The pH peaked at Day 2 (8.45) and again on Day 54 (8.68). The absence of resazurin dye color change in the surrogate flask verified that the test system remained anaerobic throughout the incubation period. Oxygen and redox potential measurements were consistent with this.

Plant Metabolism

In a plant metabolism study (6) green beans, strawberries, and beets grown in soil treated with [^{14}C]chloropicrin at a rate of 500 lb/acre were analyzed for total radioactivity and the presence of [^{14}C]metabolites. The field portion of the study was conducted by PTRL West (Richmond, CA) and Plant Sciences (Watsonville, CA) at the Watsonville facility. Plant extraction, radiocarbon analyses and characterization work was performed at PTRL East (Richmond, KY).

Test System. [^{14}C]Chloropicrin (97.8%) was obtained from New England Nuclear with a specific activity of 7.78 mCi/mmol. Reagents were obtained from Fisher Scientific (Pittsburgh, PA), Eastman Kodak (Rochester, NY) and Sigma Chemical (St.Louis, MO). All solvents were obtained from Fisher Scientific and were HPLC grade or better. Alpha-amylase (Type I-A, DFP or PMSF treated, porcine pancreas) and Pronase E (Type XXV, *Streptomyces griseus*) were obtained from Sigma Chemical. The soil in the test plot was classified as silt loam.

Six plastic casings, 12 inches in diameter, 24 inches long and open at each end, were inserted vertically into the ground such that the upper end protruded slightly above ground level.   Each casing had an area of 0.7854 ft^2 and was covered with a commercial polyethylene tarp prior to dosing.

Dosing was performed on April 24, 1989 with each of the three [^{14}C]chloropicrin treated casings receiving 2.5 mL (16.4 mCi) injected through the plastic tarp. Individual glass syringes equipped with 18 gauge stainless steel needles were used to puncture the tarp and deliver the dose six inches below the soil surface. Upon withdrawal of the emptied syringe, the puncture hole was immediately covered with a square piece of tape. The other three casings were dosed in an identical manner with unlabeled chloropicrin to serve as check plots. Two days after treatment, the tarps were removed. This injection depth and tarping interval reflects commercial application conditions for chloropicrin.

Fourteen days after application of the chloropicrin, green bean seeds, red beet seeds and bare root strawberry plants were planted, one crop per casing, in the radiolabeled and control plots respectively. Eight strawberry plants (Muir; Plant Sciences, Inc., Watsonville, CA), 16 bean seeds (Blue Lake; H. Lilly, Portland, OR) and 24 beet seeds (Detroit Dark Red; H. Lilly, Portland, OR) were planted. Irrigation water, pesticide applications (Sevin and Diazinon) and fertilizer applications (Peters 20/20/20) were made as required throughout the study.

Sampling and Analysis.   Stem, leaves, roots and fruit were harvested 66 days after planting and at maturity. Soil samples were taken at varying levels down to 24 inches, 2, 14, and 70 days after dosing and at harvest. Harvested crop and soil samples were shipped on dry ice to the laboratory within 24 hours of collection.

During the characterization period, all samples were stored at -25^{0}C. Plant and soil samples were combusted to [^{14}C]carbon dioxide using a Packard Model 306 Biological Sample Oxidizer. LSC radioassays were accomplished with a Packard 1500 liquid scintillation spectrometer. To minimize chemiluminescence the chlorite oxidation products from the lignin fractions were mixed with Hionic-Fluor cocktail (Packard). The green bean leaf acetonitrile extracts were clarified with commercial bleach prior to LSC.

All plant components were successively extracted with sodium bicarbonate and acetonitrile. Duplicate 10 g subsamples (except for green bean pods and roots, 15 g and 2 g respectively) were weighed into 45 mL polypropylene tubes. Fifteen mL of 1N sodium bicarbonate (10 mL for strawberry fruit) were added and the tissue homogenized at room temperature with a Polytron (Brinkmann, PT 10-35) for three minutes. The homogenate was then centrifuged for 25 minutes at 4^{0}C. The supernatant was decanted and the insoluble residue was homogenized two additional times with 1N sodium bicarbonate. The extraction procedure was then repeated with acetonitrile. In all cases the sodium bicarbonate removed the majority of the extractable radiocarbon. Each of the six extraction solutions were weighed and radioassayed by LSC. The insoluble residues were dried, weighed, and combusted to quantify the bound radiocarbon.

HPLC analyses of the sodium bicarbonate extracts revealed that the radiocarbon was comprised exclusively of polar compounds eluting with the solvent front in the reverse phase system. The HPLC system used a Perkin Elmer Series 4 LC equipped with a Supelco C18 column (25 cm x 0.46 cm) and an LC 90 UV detector (210nm) or a Spectra-Physics SP 8700 XR ternary pump also equipped with a Supelco C18 column. The SP 8700 was equipped with a Micromeretics 787 Variable UV/Vis detector (210nm) and a Radiomatic Flo-One Beta radioactive flow detector configured for the sequential generation of a UV absorbance spectrum and a radiochromatogram. Radiochromatograms were reconstructed using a computer program developed by the laboratory. Both HPLC units utilized a linearly programmed gradient (1ml/min) of 100% water from 0-5 min; 100% water to 100% acetonitrile from 5-25 min; 100% acetonitrile from 25-30 min; 100% acetonitrile to 100% water from 30-35 min; 100% water from 35-40 min. The effluent was fractionated into LSC vials with an ISCO fraction collector. Radiochromatograms were then reconstructed using a computer program developed by the laboratory.

A standard was prepared containing sodium bicarbonate, nitromethane, and chloropicrin. Nitromethane was included in the analysis since it had been identified as an intermediate product in the degradation of chloropicrin to carbon dioxide. The radiochromatograms for the sodium bicarbonate extracts of all three crops revealed only a single peak with a retention time of ~4 minutes. No nitromethane or chloropicrin peaks were present. Since the solubilities of these two compounds are relatively low (0.16% and 10.5% respectively at 20^0C) and the radiocarbon eluted during the 100% water phase, it was apparent that the radiocarbon was water soluble and polar in nature.

Further efforts to identify the radiocarbon in the sodium bicarbonate extract were made utilizing thin layer chromatography. Analyses of the green bean leaves, strawberry fruit and beet fruit were conducted using a polar solvent system (butanol:acetic acid:water, 4:1:1, v:v:v). Essentially all of the radiocarbon remained at the origin, further verifying the polar nature of the radiocarbon in the sodium bicarbonate extract.

A third method of analysis, descending paper chromatography, was then used to obtain separation and tentative identification of the radiocarbon. Because of the high radioactivity necessary for this type of analysis and since the HPLC chromatograms were virtually identical for all sample extracts, the strawberry fruit extract was chosen as being representative of the three crops. Descending paper chromatography analysis was performed utilizing Whatman 3 mm Chr machine direction paper eluted with n-butanol: ethanol: water (3:0.67:1, v:v:v). After drying, 1 cm wide strips were eluted with 33% methanol and quantitated by LSC. Since it was postulated that the polar ^{14}C-containing compounds were carbohydrates arising from incorporation of the radiolabeled carbon atom via the single carbon pool, a standard solution containing [^{14}C]glucose, [^{14}C]fructose, [^{14}C]sucrose, [^{14}C] starch, and [^{14}C]citrate (New England Nuclear) was used to obtain R_f values which corresponded to sucrose, fructose and glucose in the extract. Starch and citrate did not move from the origin.

The strawberry fruit sodium bicarbonate extract was further characterized by separating the ethanol insoluble, pyridine soluble, anionic, cationic and neutral fractions. The extract (49.8 mL) was added to 150 mL absolute ethanol and the precipitate (macromolecules, i.e. proteins) was removed by centrifugation. The supernatant was filtered into a round bottom flask and the extract was concentrated by rotary evaporation to dryness. The remaining residue was dissolved in pyridine and was heated over steam for ten minutes. The sample was then filtered and rinsed with an additional 2.5 mL of pyridine after which it became viscous, indicating the presence of carbohydrates. The majority of the radiocarbon was contained in this fraction. Separation of the anionic residues was accomplished using an ion exchange column eluted with 1N HCl. Neutral and cationic residues were separated using a cation exchange column eluted with water, then 1.5N NH_4OH.

The bound radiocarbon in the extracted plant tissue was then characterized via a cell wall fractionation scheme (7)(8) to separate starch, protein, pectin, lignin, hemicellulose and cellulose.   An initial wash of ~1 g of the dried residue was accomplished with 100 mL of 50mM potassium phosphate pH 7 buffer.

The filter cake was then hydrolyzed using α-amylase (20 hours, 30°C), and filtered to isolate the starch fraction. The starch-extracted pellet was then suspended in a Tris-HCl buffer (pH 7.2), hydrolyzed using pronase E (16 hours, 30°C), and filtered. The resulting filtrate was radioassayed as the protein fraction. The protein-extracted pellet was then extracted with an EDTA/sodium acetate pH 4.5 buffer (6 hours, 80°C) and filtered. The resulting filter cake, containing pectin, was washed with water and air dried. The solid plant residue was then treated three times with glacial acetic acid and sodium chlorite 2-hydrate to remove lignin. The delignified residue was then incubated in 24% KOH for 24 hours (27°C), adjusted to pH 4.5 with acetic acid and centrifuged. The supernatant contained the hemicellulose fraction.   The remaining pellet was suspended in 72% sulfuric acid for 4 hours at room temperature before being neutralized with KOH. The potassium sulfate was then dissolved in water and filtered to obtain the cellulose fraction.

Radiocarbon Distribution. Total radiocarbon in the soil declined substantially during the study in the 0-6 inch layer, dropping from 35-50 ppm two days after dosing to 5-10 ppm by the time the crops were harvested. Radiocarbon levels declined with increasing soil depth for all sampling intervals with the 18-24 inch layer containing only 2-4 ppm two days after dosing and 1.3 ppm at harvest.

Total radiocarbon in the immature plants harvested after 66 days ranged from 1.9 to 8.4 ppm for all components. At maturity, levels declined substantially, ranging from 0.1 to 2.7 ppm. Of the radiocarbon present in all plant components at maturity, 17.7 - 60.7% of the total was extracted with 1N sodium bicarbonate. For the strawberry fruit, beet and bean pods respectively, 53.7, 60.7 and 58.9% of the total radiocarbon was extracted with sodium bicarbonate.   An additional 1.3 - 6.1% was extracted with acetonitrile.

The remainder of the radiocarbon was associated with the bound fractions. With the exception of green bean leaf, every bound fraction contained radiocarbon in the starch, protein, pectin, lignin, hemi-cellulose and cellulose fractions. Levels in the strawberry fruit, leaf, stem and root were 0.007-0.578 ppm; beet root/hypocotyl, leaf and stem, 0.003-0.140 ppm; green bean pods, leaf, stem and root, ND-0.307 ppm. The analytical limit of detection for plant samples was 0.002-0.02 ppm depending on the sample size. Total radiocarbon recovery for all plant analyses was $101.2 \pm 10.0\%$.

Plant Metabolites. No chloropicrin or nitromethane was detected in any of the bound or extractable fractions. HPLC and TLC characterizations indicated the polar nature of the radiocarbon in the sodium bicarbonate extracts. Paper chromatographic analysis tentatively identified the sodium bicarbonate extractable radiocarbon as glucose, fructose, and sucrose.   The remaining bound ^{14}C was characterized by cell fractionation procedures and found to be distributed among starch, protein, pectin, lignin, hemicellulose, and cellulose.

Of the strawberry fruit sodium bicarbonate extract, 30.8% of the total radiocarbon was present in the pyridine soluble fraction and 22.4% was tentatively identified as citrate, glucose, fructose and/or sucrose. The ethanol insoluble fraction of the sodium bicarbonate extract contained 4.5% of the total radiocarbon, indicating the presence of ^{14}C-proteins. Of the total radiocarbon present in the anionic (ND), cationic (4.0%) and neutral fractions (17.3%), the largest single component was tentatively identified as ^{14}C-glucose, ^{14}C-fructose, and/or ^{14}C-sucrose.

These results indicate that the radiocarbon present in plants grown in soil treated with [^{14}C]chloropicrin consists of natural plant biochemical components which have incorporated the radiolabeled carbon atom. Since carbon dioxide has been identified as the terminal breakdown product of chloropicrin in soil under aerobic conditions([1]), incorporation of [^{14}C]carbon dioxide via the single carbon pool is an obvious mechanism for this. Furthermore, the major concentrations of radiocarbon in each plant tissue were generally associated with the natural plant products present in the tissue. For example, 32-35% of the radiocarbon found in stem tissue for all three crops was associated with lignin, hemicellulose, and cellulose. The uptake of carbon dioxide by plant roots and its metabolism into numerous endogenous products has been well documented in the published literature ([9]).

Photohydrolysis

In a study ([10]) conducted at Bolsa Research Associates (Hollister, CA), the half-life of chloropicrin in photolyzed water at 25°C was determined. Chloropicrin dissolved in water was irradiated with a xenon lamp and the nature and extent of degradation products formation was examined under simulated sunlight conditions.

Test System. Unlabeled chloropicrin (99.7%) was dissolved in nanograde water (Type II / Grade II) to a concentration of 800ppm. All water was deionized with a commercial unit, purified with a Barnstead/Thermolyne cartridge and sterilized. A pH buffer was added to the water prior to introduction of the chloropicrin.

The buffer solution was prepared by mixing 250 mL of 0.100 M potassium dihydrogen phosphate and 145 mL of 0.100 M sodium hydroxide and adjusting the volume to 500 mL. The solution was then autoclaved.

The 800 ppm chloropicrin stock solution was then diluted to 164 ppm (0.001 M) with buffered water. This 0.001M test solution was then added by syringe to 103 Kimble 12 mL vials with pressure sealed teflon lined septum caps. The vials were filled such that there was no observable headspace.

Three vials taken at random were analyzed initially to validate the Time 0 concentration. Fifty were covered with foil, inverted, and stored in the dark in a 25.0°C circulating water bath to serve as controls. The remaining 50 vials were then inverted and placed in a circulating water bath thermostatted at 25.0°C. Approximately half of each vial (3.5 cm) protruded out of the water for exposure to the light source.

The irradiation source was a Suntest CPS (Heraeus Co.) photomachine with a xenon lamp. The light intensity in the photochamber was set at 1100 lux (lumens/m^2) as measured by a Extech Instruments photometer (#3-13895/590). During the ten days of photoexposure (April 3 - 12, 1993) the maximum intensity of outdoor sunlight at the Hollister, California, test site was measured with the same photometer and found to be approximately 50% of the xenon lamp intensity.

Three chloropicrin standard solutions were prepared (0.0001M, 0.0005M, 0.001M) in the pH 7.0 buffer. Carbon dioxide standards (3.2, 6.3 and 25.5 ppm) were prepared in 6 L Luxfer aluminum cylinders by the partial pressure method. These were calibrated against a 51.0 ppm certified gas mix (Scott Speciality Gases; Plumstead, PA). Inorganic chloride standards (100, 50, 10, and 1 ppm) were prepared from a 1000 ppm stock solution obtained from Banco (IC No. 18250). The inorganic nitrate standards (100, 50, 10, and 1 ppm) were prepared from a 1000 ppm stock solution (Banco IC No. 44130). Inorganic nitrate standards (0.02, 0.04, 0.08, and 0.2 mg/L) were prepared by dissolving dried anhydrous sodium nitrite in distilled water.

Sampling and Analysis. The test solutions were irradiated in the photochamber for twelve continuous hours daily. Sampling was performed after 12, 24, 36, 48, 60, 72, 84, and 108 hours of irradiation. At each sampling interval 3-5 vials were analyzed for chloropicrin, carbon dioxide, inorganic chloride, nitrate, nitrite and pH. All analyses were performed within eight hours of sampling.

Chloropicrin levels were quantified by gas chromatography with a HP 5890 Series II GC equipped with a J&W Scientific DB624 megabore column (30 m x 0.53 mm) and flame ionization detector. A Corning Model 250 pH/Ion Analyzer was used to measure pH (Orion model 90-02 electrode), nitrate (Orion model 93-07 electrode), and chloride (Orion model 94-17B) concentrations. Carbon dioxide levels were determined by direct injection into the GC/MS. Bicarbonate ion concentrations were calculated from the measured levels of carbon dioxide using the relationship between free CO_2, HCO_3^-, CO_3^{2-} and pH (*11*). At a pH of 7.0, for example, a 20% concentration of carbon dioxide is in equilibrium with 80% bicarbonate.

Photolytic Half-life. When exposed to a xenon light source, a 0.001M chloropicrin solution maintained at 25^0C and pH 7 degraded quantitatively to carbon dioxide (bicarbonate, carbonate), chloride, nitrate, and nitrite. No other volatile photoproducts were detected during GC analysis. Assuming pseudo-first order kinetics, the linear regression analysis of the plot of ln[chloropicrin] vs time for the eight sampling intervals derived a photolytic half-life of 31.1 hours ($r^2 = 0.9506$) . The photolysis of chloropicrin proceeded as shown in Figure 3.

$$CCl_3NO_2 \xrightarrow[H_2O]{h\upsilon} 3Cl^- + NO_2^- + NO_3^- + CO_2\,(HCO_3^- + CO_3^{2-}) + H^+$$

Figure 3. Photolysis of Chloropicrin

The average recovery was $90 \pm 16\%$ for all samples based on the measurements of chloropicrin and carbon dioxide. Individual recoveries based on the chloride, nitrite plus nitrate, and carbon dioxide measurements averaged 104.5% of the nominal dose at Time 0, decreased to a low of 68.9% at 48 hours and increased to 80.9% at the 108 hour interval. The low recoveries may have been due to sampling procedures during the intermediate times.

There was no measurable hydrolysis of chloropicrin in the dark controls over the ten day study period. This confirmed that the important process for the degradation of chloropicrin in water was photohydrolysis.

Summary

The half-life of [^{14}C]chloropicrin in sandy loam soil dosed at a rate equivalent to 500 lb/acre and incubated at 25^0 C in the dark was 4.5 days. Chloronitromethane and nitromethane, two transient intermediates in the aerobic metabolism of chloropicrin, were identified. After 24 days, up to 75% of the applied radiocarbon was present as [^{14}C]carbon dioxide and 14.7% was bound in the soil.

A mixture of soil flooded with water and dosed with [^{14}C]chloropicrin at a rate equivalent to 500 lb/acre was incubated at 25^0 C in the dark under anaerobic conditions. The chloropicrin degraded readily to nitromethane with a half-life of 1.3 hours. Chloronitromethane was present at 80.3% of the dose at the Time 0 sampling.

In a plant metabolism study, green beans, strawberries, and beets grown in soil treated with [^{14}C]chloropicrin at a rate of 500 lb/acre were analyzed for total radioactivity and the presence of [^{14}C]metabolites. No chloropicrin or nitromethane was detected in any of the bound or extractable fractions. Chromatographic analyses tentatively identified the sodium bicarbonate extractable radiocarbon as glucose, fructose, and sucrose. The remaining bound ^{14}C was characterized by cell fractionation procedures and was found to be distributed among starch, protein, pectin, lignin, hemicellulose, and cellulose. Since carbon dioxide has been identified as the terminal breakdown product of chloropicrin in soil under aerobic conditions, an obvious mechanism for incorporation of the radiocarbon is by metabolism of [^{14}C]carbon dioxide via the single carbon pool.

When exposed to a xenon light source, a 0.001M chloropicrin aqueous solution maintained at 25^0C and pH 7 degraded quantitatively to carbon dioxide (bicarbonate, carbonate), chloride, nitrate, and nitrite. No other volatile photoproducts were detected. The half-life of chloropicrin was 31.1 hours. In the absence of light, chloropicrin did not undergo hydrolysis.

Acknowledgments

These studies were funded by the Class I members of the Chloropicrin Manufacturers Task Force: Ashta Chemicals Inc. (Ashtabula, OH), Holtrachem Mfg. (Orrington ME), Niklor Chemical Co. (Long Beach, CA) and Trinity Mfg. (Hamlet, NC). Technical assistance was provided by John Butala, AgrEvo Canada (Regina, Saskatchewan), Angus Chemical Co. (Buffalo Grove, IL), DowElanco (Indianapolis, IN), Great Lakes Chemical Corp. (West Lafayette, IN), and TriCal Inc. (Hollister, CA).

Literature Cited

(1)   Shepler, K.; Hatton, C. *Aerobic Soil Metabolism of [^{14}C]Chloropicrin*; PTRL West Inc. Project 448W, 1995, unpublished report.
(2)   Skinner, W.; Jao, N. *Laboratory Volatility of [^{14}C]Chloropicrin*; PTRL West Inc. Project 449W, 1995, unpublished report.

(*3*)   Toia, R.F.; Sprinkle, R.B. *Development of Extraction and Analysis Methods Preliminary to a Confined Rotational Crop Study with [^{14}C]Chloropicrin Using Lettuce, Carrot, Wheat and Strawberry;* PTRL West Inc. Project 501W, 1995, unpublished report.

(*4*)   Shepler, K.; Hatton, C. *Anaerobic Aquatic Metabolism of [^{14}C]Chloropicrin;* PTRL West Inc. Project 448W, 1995, unpublished report.

(*5*)   Castro, C.E.; Wade, R.S. and Belser, N.O. *J. Agric. Food Chem.*, 1983, 31, pp. 1184-1187.

(*6*)   Lawrence, J. *Quantitative Characterization of [^{14}C]Residues Present in Soil, Strawberries, Green Beans and Red Beets Grown Under Actual Field Conditions Following Treatment of Soil with [^{14}C]Chloropicrin;* PTRL East Inc. Project 327, 1990, unpublished report.

(*7*)   Kovacs, M.J.; *Residue Reviews*, 1986, *97*, pp. 1-17.

(*8*)   Langebartels, C.; Harms, H. *Ecotoxicol. and Environ. Safety*, 1985, pp. 268-279.

(*9*)   Basra, A.S.; Malik, C.P. *Biol. Rev.,60*, 1985, pp. 357-401.

(*10*)   Lee, H.; Moreno, T.; *Photohydrolysis of Chloropicrin*, Bolsa Research Assoc. Project BR389.1:93, 1993, unpublished report.

(*11*)   Schmitt, C. *Annls. Ec. Natu. Sup. Mecanique*, Nantes, 1955.

Chapter 9

Reducing 1,3-Dichloropropene Air Emissions in Hawaii Pineapple with Modified Application Methods

Randi C. Schneider[1], Richard E. Green[1], Calvin H. Oda[3],
Brent S. Sipes[2], and Donald P. Schmitt[2]

[1]Department of Agronomy and Soil Science, University of Hawaii,
1910 East West Road, Honolulu, HI 96822
[2]Department of Plant Pathology, University of Hawaii,
3190 Maile Way, Honolulu, HI 96822
[3]Del Monte Fresh Product (Hawaii), Kunia, HI 96759

Restrictions on 1,3-dichloropropene (1,3-D) use in California due to air quality concerns prompted the testing of improved application methods in Hawaii. Strategies investigated for their impact on 1,3-D air emissions include (1) a comparison of one and two soil chisels per bed, (2) polyethylene mulch films which cover one or two beds, and (3) an emulsified liquid formulation of 1,3-D (SL) applied by drip irrigation compared with chisel injected 1,3-D. Air concentrations were measured at a 15-cm height above the soil to compare 1,3-D emissions near the source. Measurements of spatial and temporal distributions of 1,3-D in soil gas and in soil profiles complemented air monitoring. A single chisel per bed (45 cm depth) reduced 1,3-D air emissions compared with double chisels (30 cm depth). Wide mulch film did not reduce air emissions in a single field trial. Although drip irrigation application of 1,3-D resulted in reduced air emissions, 1,3-D soil distribution was sub-optimal compared with chisel injection.

Preplant soil fumigation is widely practiced in Hawaii's pineapple industry to control plant parasitic nematodes in soil (1,2). At present, 1,3-dichloropropene (1,3-D) and methyl bromide are the two soil fumigants in widespread use. Recent legislation phasing out the

0097–6156/96/0652–0094$15.00/0
© 1996 American Chemical Society

use of methyl bromide by the year 2001 (*3*), has prompted pineapple companies to evaluate alternative fumigants, as well as non-volatile nematicides for preplant nematode control. In 1990, the use of 1,3-D was restricted in California due to potential negative impacts on the air quality during spring fumigations (*4*). Regulatory action in California prompted the pineapple industry to evaluate air emissions resulting from the current use patterns of 1,3-D in Hawaii (Hawaii Department of Agriculture, unpublished report). The industry began testing reduced application rates of 1,3-D and new application methods to lower air emissions to acceptable levels.

Pineapple growers have typically applied 1,3-D at a rate of 224 to 336 L ha^{-1}. Pineapple was grown in a perennial cropping cycle which yields three fruit harvests from a single planting. In addition, fields were fallowed for 6-12 months after the third harvest. Therefore, each field was fumigated once every 4-5 years. 1,3-D was typically applied by chisel injection to a depth of 30-40 cm, with two chisels spaced 46 cm apart, to correspond to two plant rows per bed. Polyethylene mulch film and drip irrigation tubing were also laid down during the fumigation process.

The objective of our research program was to evaluate the impact of three new application methods on 1,3-D air emissions above treated fields. In each experiment, a new application method was compared with a conventional method by sampling 1,3-D in air at a fixed height (15 cm) above the field within the treated areas. Collecting 1,3-D air samples at a single height allowed us to directly compare emissions between treatments but did not allow us to quantify the flux of 1,3-D from the treated fields. The application methods which were evaluated in four large field experiments were: 1) single chisel injection compared with double chisel injection; 2) wide polyethylene mulch film (2 m wide) compared with narrow polyethylene mulch (0.8 m); 3) drip irrigation application of 1.3-D compared with chisel injection. The single and double chisel injection treatments were evaluated in two field experiments, in 1990 and 1993.

Materials and Methods

Field Experiments.
Field experiments were conducted near Kunia, Oahu, Hawaii in commercial pineapple fields on Del Monte plantation. The field sites are summarized in Table I. Telone II (94 % A.I. 1,3-D, DowElanco) was applied in all four experiments at an application rate of 224 L ha^{-1} using commercial fumigation equipment. Two pineapple beds were fumigated simultaneously with chisels mounted on a tractor which formed the bed, laid polyethylene mulch film and drip irrigation tubing, and applied granular fertilizer during the fumigation operation. In experiments I, II, and III, the soil series was Wahiawa silty clay, an aggregated, well drained Oxisol with a bulk density of 1.0 g cm^{-3}, and an organic carbon content of 1.5 to 2 %. In experiment IV, the soil series was Kunia silty clay, an Inceptisol with similar soil properties, but without stable aggregates. Soil aggregates result in a coarse textured soil with drainage properties similar to sandy soils.

In experiment I, two different chisel injection methods were compared for their effect on air emissions of 1,3-D. In the single chisel treatment, fumigant was injected in the center of the bed at a depth of 45 cm. The double chisel method used two chisels per bed, spaced 46 cm apart, with an injection depth of 35-40 cm. In addition to the chisel treatments, polyethylene mulch film (0.8 m wide, 1 mil thick) was used in treatments 1 and 2, but not treatment 3 (Table I). Treatment blocks varied in size from 0.4 to 3.9 ha and were separated by untreated buffer zones to minimize cross contamination during air monitoring. Field blocks were fumigated with 1,3-D on December 7, 8, and 10, 1990 for treatments 1, 2, and 3, respectively. Three air sampling pumps were centrally located in each treatment at a sampling height of 15 cm above the soil surface. Air samples were collected in 12-hour sampling intervals for 5 days, followed by 24-hour intervals for the remainder of the 10-day sampling period.

Table I. Summary of field experiments used in 1,3-D air monitoring study

Expt.	Treatments	Date (Field)	1,3-D Rates (L ha^{-1} A.I.)	Soil Order	1,3-D Measurements
I	1. Double chisel + mulch 2. Single chisel + mulch 3. Single chisel, no mulch	Dec. 1990 (DM-1)	224	Oxisol	Air Soil gas
II	1. Single chisel 2. Double chisel	August 1993 (DM-13)	224	Oxisol	Air, Soil gas Soil
III	1. Narrow mulch 2. Wide mulch	Sept. 1993 (DM-13)	224	Oxisol	Air, Soil gas Soil
IV	1. Drip irrigation 2. Single chisel	August 1994 (DM-5)	224	Inceptisol	Air, Soil gas Soil

In experiment II, single and double chisel injection methods were also compared for their effect on 1,3-D air emissions. A new single chisel injector design, a split-depth injector, was used. Two treatment blocks (block size 2.7 ha) separated by an untreated buffer zone (266 m wide) were fumigated with 1,3-D using either single chisel injection or the conventional double chisel method. With the split-depth single chisel injector, 70 % of the fumigant was injected at 45 cm depth and the remaining 30 % injected at 30 cm. Double chisel injection delivered the fumigant to 40 cm depth (chisel spacing of 46 cm). Polyethylene mulch film (0.8 m wide, 1 mil) was used in both treatments. Fumigation of both treatments was accomplished on the same day, August 17, 1993. Two air monitoring stations were established in the center of each treatment block. The air sampling height was 15 cm and the sampling interval was 12-hours during the entire 7-day sampling period.

Experiment III was conducted in September 1993, in the same field as experiment II, to compare two types of polyethylene mulch for their retention of 1,3-D. Two adjacent field blocks (separated by a road, 5 m wide) were fumigated with 1,3-D using single chisel split-depth injection. Treatment 1 was covered with standard narrow polyethylene mulch (0.8 m) and treatment 2 was fumigated using wide mulch (2 m) to cover two planting beds; both mulch films were 1 mil in thickness. Field blocks were fumigated on September 14 and 15, for wide mulch, and narrow mulch treatments, respectively. Two air sampling stations per treatment and 12-hour sampling intervals were used.

In experiment IV, drip irrigation application of 1,3-D SL, an emulsifiable formulation, (XRM-5053, DowElanco, 66 % A.I. 1,3-D) was compared with single chisel injection of 1,3-D (Telone II, 94 % A.I. 1.3-D) using a new chisel design, a winged shank injector developed by DowElanco. Two field blocks, 0.5 and 1.4 ha in size, were fumigated by drip irrigation and single chisel injection, respectively. Polyethylene mulch film (0.8 m wide, 1 mil) was used in both treatments. The two treatment blocks were separated by a buffer zone (365 m) and five air monitors were installed in each treated block, as well as in the center of the untreated zone. 1,3-D SL was applied with 1.9 cm of water, continuously during a 6-hour irrigation cycle. 1,3-D was injected at 45 cm depth in the center of the bed, using the winged-shank injector. The two wings on the injector were offset from the chisel shaft by 7.5 cm on each side; this design minimizes the upward loss of 1,3-D through the chisel trace. Both 1,3-D treatments were applied on August 23, 1994 and 1,3-D air emissions were monitored for 14 days. A 12-hour sampling interval was used for the first 7 days followed by a 24-hour interval thereafter.

Analytical Methods

Air Sampling Methods. Air samples for 1,3-D were collected using SKC air monitoring pumps (Model 224) and coconut charcoal adsorbent tubes (SKC, 10 mm diameter, 800/200 mg) at a pumping rate of 1.25 liters per minute. Pumps were supplied with charged batteries and calibrated with a flow meter at the start of each sampling period. Sample tubes were stored in a freezer soon after collection. Field spikes and blank tubes were included with each day's samples. Air samples from experiments I and IV were analyzed by Dow Chemical Co. using solvent desorption with carbon disulfide and GC-FID or GC-ECD (Dow Chemical Co., unpublished report). Air samples from experiments II and III were analyzed by solvent desorption with acetone and analyzed by GC-ECD (5). Charcoal samples were sonicated for 30 minutes in 10 ml of acetone and analyzed by direct injection on a HP5890A gas chromatograph with HP7673A autosampler. Extracts were analyzed using isothermal conditions (70 C) and a DB-1701 column (15 m x 0.53 mm, 1 μm film, J.& W. Scientific). Retention times for cis and trans isomers of 1,3-D were 2.17 and 2.65 minutes, respectively. In laboratory spikes, sample tubes were fortified with 5 μl of Telone II standard (DowElanco, 96 % 1,3-D; cis : trans ratio, 53 : 47) to yield concentrations of 50.7 and 45.1 μg per tube, for cis and trans 1,3-D, respectively. Recovery of 1,3-D averaged 90.6 % (± 4.6) for cis 1,3-D and 93.9 % (± 4.6) for trans 1,3-D. Recovery of 1,3-D from field spikes averaged 81.2 % (± 3.1) and 80.2 % (± 5.1), for cis and trans isomers, respectively. Samples were frozen and analyzed within two weeks of collection. A storage stability study showed no loss of extraction efficiency over a 2-week period.

Soil Gas Method. Soil gas samples were collected and analyzed for 1,3-D during the first seven days after fumigation. Gas samples were collected in two locations, the planting row (23 cm from the bed center), and the interbed (53 cm from bed center). Stainless steel probes (1/8 inch OD) were used to collect gas samples at depths of 5, 15, 30, and 45 cm (6-7). Soil probes were sealed with Swagelok unions and rubber septa. Gas samples were collected with 0.5 ml gas sampling syringes (Dynatech A-2) after purging the probe volume by removing 3-ml of air from each probe. Gas samples were stored at room temperature and analyzed within 30 hours of collection by GC-ECD as described for air samples. Soil gas samples (50 μl) were injected manually and cis and trans 1,3-D were quantified with gas standards made from a Telone II standard.

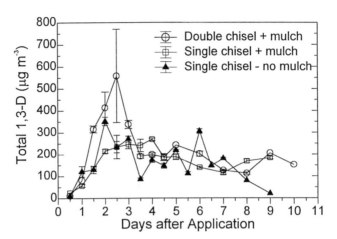

Figure 1. Concentrations of total 1,3-D in air (cis + trans, μg m⁻³) for experiment I are plotted over a 10-day period after fumigation. Values are means of triplicate samples with standard error bars. (Reproduced with permission from ref. 7. Copyright 1995 Butterworth-Heinemann journals, Elsevier Science Ltd, The Boulevard, Langford Lane, Kedlington 0X5, UK.)

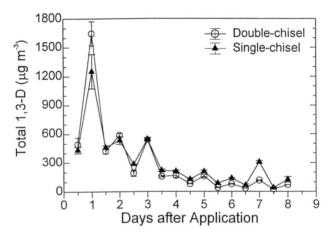

Figure 2. Concentrations of total 1,3-D in air (cis + trans, μg m⁻³) for experiment II are plotted over an 8-day period. Values are means of duplicate samples with standard error bars. (Reproduced with permission from ref. 7. Copyright 1995 Butterworth-Heinemann journals, Elsevier Science Ltd, The Boulevard, Langford Lane, Kedlington 0X5, UK.)

Soil Sampling Method. Soil samples were collected 7-8 days after fumigation with bucket augers to a maximum soil depth of 90 cm, in 15-cm increments (*6-7*). Soil profiles were sampled from bed center, planting row, and interbed to describe the spatial distribution of 1,3-D in the bed. Soil samples were stored at 4 C for up to four days after collection. Subsamples (50 g) were extracted by co-distillation, using a mixture of 175 ml water and 10 ml hexane (*8*). Sample extracts were analyzed by GC-ECD as described for air samples. In laboratory spikes, soil samples fortified at a concentration of 0.04 $\mu g\ g^{-1}$, yielded average recoveries of 90.3 % (± 7.2) and 82.1 % (± 9.6), for cis and trans 1,3-D, respectively.

Results

In experiment I, conducted in December 1990, 1,3-D air emissions from double chisel fumigation were compared with emissions from two single chisel treatments, with and without mulch film. The target injection depths were 45 cm for the single chisel, and 35-40 cm for the double chisel. The air monitoring results, reported in Figure 1, show much higher 1,3-D air emissions from double chisel fumigation compared to the single chisel. There was little difference in the air emission pattern between the two single chisel treatments. The polyethylene mulch had little or no effect in retaining the fumigant.

In experiment II, conducted in August 1993, a new single chisel design, the split-depth injector, was compared with the double chisel injection method. 1,3-D air emissions for experiment II are plotted in Figure 2. Peak values measured during this experiment were two to three times higher than those measured in experiment I (Figure 1). Both chisel treatments showed similar 1,3-D air concentrations with the exception of the peak values measured 24 hours after fumigation.

The design of experiment III, conducted in September 1993, incorporated split-depth single chisel injection and two types of mulch film, (narrow and wide) to assess the effect of mulch type on 1,3-D air emissions. The 1,3-D air emission pattern measured in this experiment, shown in Figure 3, was quite different from the two previous field trials. There was a large diurnal fluctuation in 1,3-D air concentration, probably as a result of unusual weather conditions. The measured 1,3-D air concentrations showed no treatment differences with the two types of mulch. Peak 1,3-D values, measured 48 hours after fumigation were slightly higher in the wide mulch treatment. Soil gas results from the planting row, plotted in Figure 4, offer an explanation for the lack of treatment differences in the air data. 1,3-D concentrations in soil gas were approximately three times higher in the narrow mulch block than in the wide mulch block. In Figure 4a, 1,3-D soil gas concentrations peaked three days after injection near the targeted split-injection depths, 30 and 40 cm. In the wide mulch block (Figure 4b) peak soil gas levels were measured earlier, two days after injection, at 5- and 15-cm depth. Very low 1,3-D concentrations were detected at 30 and 40 cm (Figure 4b). These data indicate that 1,3-D did not penetrate to the target injection depths in the wide mulch block. It was later confirmed that soil tillage was poor in the wide mulch block, preventing deep penetration of the chisels.

Air emission data from experiment IV, which compared drip irrigation of 1,3-D SL, with chisel-injected 1,3-D, using a winged shank, are plotted in Figure 5. Air monitoring during the first 48 hours after application showed higher 1,3-D emissions from the drip irrigation treatment. The initial air samples, plotted at Day 0, represent a 3-hour interval immediately after application (1530-1830 hours). All other data points represent 12- or 24-hour intervals. The initial differences in 1,3-D air emissions between treatments are

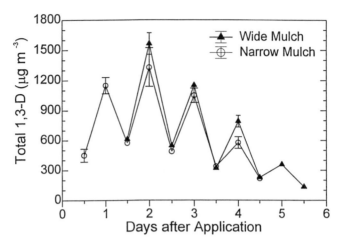

Figure 3. Concentrations of total 1,3-D in air samples (cis + trans, μg m⁻³) from experiment III are plotted over a 6-day period after fumigation. Values are means of duplicate samples with standard error bars.

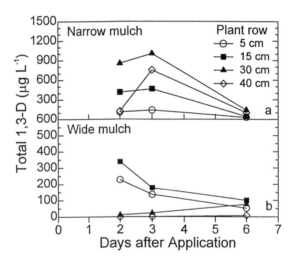

Figure 4. Soil gas concentrations of total 1,3-D (cis + trans, μg L⁻¹) from experiment III collected in the planting row from 4 soil depths. Values are means of 4 replicate samples. a) Treatment 1, single chisel injection with narrow (0.8 m) polyethylene mulch. b) Treatment 2, single chisel injection with wide (2 m) mulch.

attributed to application method. With chisel injection, the fumigant was delivered in a narrow band at 45 cm depth. With drip application, 1,3-D SL was applied continuously with 1.9 cm of water over a 6-hour irrigation period. Irrigation resulted in a wider initial distribution of 1,3-D both vertically and laterally in the planting bed. The soil water content also differed significantly between treatments. The low soil water content (15-25 %) in the chisel treatment represents a soil moisture tension of approximately 100 kPa, compared with 28-33 %, or 10-30 kPa in the irrigated treatment. Soil gas 1,3-D concentrations from the interbed location are plotted in Figure 6. Gas concentrations were relatively low (< 100 μg L^{-1}), and represent loss of 1,3-D from the mulch-covered planting bed. Figure 6b shows a diffuse 1,3-D gas distribution in the drip irrigation treatment with peak values at 45 cm, three days after application. With chisel injection, maximum 1,3-D levels were measured at 30 cm depth, two days after fumigation (Figure 6a). Soil gas 1,3-D values at 5 cm depth were significantly higher with drip irrigation than with chisel injection. These 5-cm soil gas data correlate well with the air emission results (Figure 5) and indicate that greater lateral movement of 1,3-D with drip irrigation resulted in greater loss of 1,3-D from the bed and therefore higher initial air emissions.

Discussion

Of the application methods tested in this study for reducing 1,3-D air emissions, deep single chisel injection was the most promising. The air emission data from the four field experiments indicates that the chisel designs used in experiments I and IV, performed better than the split-depth chisel design. 1,3-D air concentrations ranged from 200 to 600 μg m^{-3} in those two experiments compared with peak values of 1200 to 1500 μg m^{-3} measured in experiments II and III, where the split-depth chisel injector was used. With single chisel injection, the distribution of 1,3-D in soil differed markedly from the double chisel method. When two chisels per bed were used, the result was uniform 1,3-D distribution in the bed, whereas the single chisel produced a sharp concentration gradient from the center to the edge of the bed (7). Field studies have shown that single chisel injection is equivalent to double chisel injection in terms of nematode control and crop yield (2,7).

Polyethylene mulch films are not usually recommended for use with 1,3-D (9), but have been used in Hawaii's pineapple industry since the initiation of 1,3-D fumigation in the 1940's (10). Plastic mulch films are used to provide weed control and conserve soil moisture as well as to retain fumigants. Since pineapple is a perennial crop and is drip irrigated, the use of mulch is a valuable component of the cropping system. In experiment I, 1,3-D air emissions were compared for deep single chisel injection, with and without mulch film. The absence of mulch had very little effect on the magnitude of 1,3-D loss to the atmosphere (Figure 1).

Wide polyethylene mulch film (2 m) has been used routinely with methyl bromide fumigation in pineapple and was tested in experiment III for its potential to reduce 1,3-D loss to the atmosphere. Due to the problems with poor soil tillage in that experiment, it was inconclusive whether wide mulch would improve 1,3-D retention. In small research plots, 1,3-D concentrations in soil gas increased along the gradient from no mulch to narrow mulch to wide mulch, indicating the value of plastic mulch in retaining 1,3-D in the soil for a longer period (Schneider, R. C., unpublished data).

Drip irrigation of 1,3-D SL compared favorably with chisel injected 1,3-D in terms of air emissions. The early peak in 1,3-D emission was due to the continuous irrigation

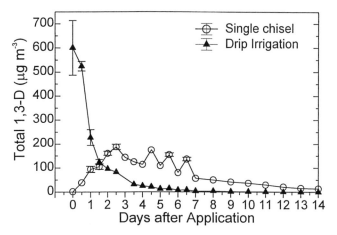

Figure 5. Concentrations of total 1,3-D in air (cis + trans, μg m⁻³) for experiment IV are plotted over a 14-day period after 1,3-D application. Values are means of 5 samples per treatment with standard error bars.

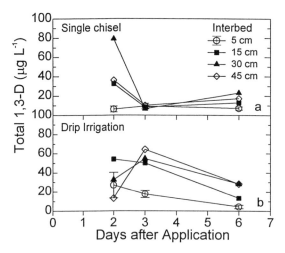

Figure 6. Soil gas concentrations of total 1,3-D (cis + trans, μg L⁻¹) from experiment IV collected in the interbed location from 4 soil depths. Values are means of 6 replicate samples, with standard error bars plotted for the 5 cm depth. a) Drip irrigation application of 1,3-D SL with 1.9 cm water. b) Single chisel injection of 1,3-D fumigant to 45 cm depth with winged shank.

method used, and the large volume of water applied. A modified drip irrigation method, which has been used previously (*6*), is to apply the fumigant formulation in a small volume of water and then post-irrigate with the desired amount of water. This method would reduce the 1,3-D concentration near the soil surface and minimize lateral transport of the fumigant in the bed. The 1.9 cm of water used in experiment IV resulted in a diffuse 1,3-D concentration in the bed, with penetration below the 45-cm pineapple rooting depth. Based on drip irrigation experiments with 1,3-D SL, and non-volatile nematicides in pineapple, (*6,11*) a 1.3 cm irrigation volume should provide a more optimal 1,3-D distribution in soil.

Acknowledgments

J. Mueller, DowElanco coordinated experiment IV and provided 1,3-D air data. We thank K. Okazaki for analytical work, and Q. Lin, C. Chen, M. Young, D. Meyer, G. Nagai, and the Del Monte research staff for field assistance.

Literature Cited

1. Albrecht, W.N., *Arch. Environ. Health* **1987**, *42*, 286-291.
2. Sipes, B.S., Schmitt, D.P., Oda, C.H., *J. Nematol.* **1993**, *25*, 773-777.
3. Braun, A.L.,. Supkoff, D.M . *Options to Methyl Bromide for the Control of Soil-Borne Diseases and Pests in California with Reference to the Netherlands.* Report PM-02 California Environmental Protection Agency. Sacramento, CA, **1994**.
4. Fitzell, D. *Ambient air monitoring in Imperial County for Telone after application to a test field by DowElanco.* Engineering Report C90-014A, Evaluation Branch, Monitoring and Lab. Div. California Air Resources Board, Sacramento, CA, **1991** .
5. Van den Berg, F., Leistra, M., Roos, A.H., Tuinstra, L.G.M. Th., *Water, Air, Soil Pollut.* **1992**, *61*, 385-396.
6. Schneider, R.C., Green, R.E., Wolt, J.D., Loh, R.K.H., Schmitt, D.P., Sipes, B.S., *Pestic. Sci.***1995**, *43*, 97-105.
7. Schneider, R.C., Sipes, B.S., Oda, C.H., Green, R.E., Schmitt, D.P. *Crop. Prot.***1995**, *14(8)*, 611-618.
8. Rains, D.M., Holder, J.W., *J. Assoc. Off. Anal. Chem.* **1981**, *64*, 1252-54.
9. Lembright, H.W., *J. Nematol.* **1990**, *22,* 632-644.
10. McFarlane, J.S., Matsuura, M., . *Phytopath.* **1947**, *37*, 39-48.
11. Schneider, R.C., Green, R.E., Apt, W.J., *Acta Horticulturae* **1993**, *334*, 351-360.

Chapter 10

Strategies for Reducing Fumigant Loss to the Atmosphere

William A. Jury[1], Yan Jin[2], Jianying Gan[1], and Thomas Gimmi[1]

[1]Department of Soil and Environmental Sciences,
University of California, 2208 Geology, Riverside, CA 92521
[2]Department of Plant and Soil Science,
University of Delaware, 149 Townsend Hall, Newark, DE 19716

A model is developed to describe transport and loss of methyl bromide (MeBr) in soil following application as a soil fumigant. The model is used to investigate the effect of soil and management factors on MeBr volatilization. Factors studied include depth of injection, soil water content, presence or absence of tarp, depth to downward barrier, and irrigation after injection. Of these factors, the most important was irrigation after injection followed by covering with the tarp, which increased the diffusive resistance of the soil and prevented early loss of MeBr. The model offers an explanation for the apparently contradictory observations of earlier field studies of MeBr volatilization from soils under different conditions. The model was also used to calculate the concentration-time index for various management alternatives, showing that the irrigation application did not make the surface soil more difficult to fumigate, except at very early times. Therefore, irrigation shows promise for reducing fumigant loss while at the same time permitting control of target organisms during fumigation.

Introduction

Current preplant field fumigation guidelines mandate that a ≥ 1 mil polyethylene film or tarp be put on the soil surface after a shallow (15–30 cm) injection of fumigant. Until recently, there was no information available about how effective the tarp was at restricting volatilization of methyl bromide (MeBr) to the atmosphere following application. In the past few years, several experimental studies of emission of MeBr from fumigated fields have been conducted to determine the amount of chemical entering the atmosphere after application. Their findings are somewhat contradictory and difficult to interpret.

The first published field scale study (1) showed that 87% of applied MeBr was lost to the atmosphere within 7 days after a commercial fumigation on a tarped field, and much of that occurred in the first few hours after application. When the same group repeated their experiment on the same soil, however, they observed a loss of only 34% of applied MeBr (2). Majewski et al. (3) conducted experiments

0097–6156/96/0652–0104$15.00/0

on adjacent fields and found 89% losses of MeBr to the atmosphere from the one that had no tarp over the surface, but smaller losses from the tarped field, i.e. 32%. Finally, Yates et al. (4) measured 64% losses from a tarped field in which MeBr was injected at the 25-30 cm depth.

A laboratory study conducted by Jin and Jury (5) demonstrated that MeBr could move readily through a 1 mil polyethylene tarp covering the soil surface, apparently by dissolving into the tarp and diffusing through it. In their system, essentially 100% of the MeBr introduced at the bottom of a 22 cm soil column volatilized out the top before degrading, even if the tarp was present. The authors were able to reduce this loss substantially by adding water before covering the surface with the tarp, to as low as 4% cumulative volatilization when 1.6 cm of water was applied.

These results imply that many factors can influence fumigant volatilization loss from soil. From the limited experimental information available, however, it is not clear what these factors are, and what relative importance they have. The purpose of this study is to employ a simplified mathematical model of the fumigation process to identify the factors influencing fumigant loss and to explore the effects of different fumigant and soil management strategies on reducing this loss. Since it will not be beneficial to reduce atmospheric loading if the fumigant is rendered ineffective at pest control in the process, we also examine the concentration-time index CT (the product of the average fumigant concentration and the time of exposure) for each scenario studied. This index has been shown in previous studies to be correlated with the efficacy of fumigant dosage (6).

Theory

Transport Equations, Initial and Boundary Conditions The scenario that will be used for the calculation of volatilization losses is as follows:

1. The soil consists of two uniform layers of different water content θ_1, θ_2. The upper layer $(0 < z < H)$ is wetter than the lower, to represent the aftermath of a recent irrigation.

2. Fumigant moves by vapor diffusion only. This is not true for the first hour or so when density and pressure gradients contribute significantly to vapor advection, but approximates the stage that initiates when the fumigant partial pressure lowers to the point where mass flow becomes less important than diffusion.

3. Loss to the atmosphere occurs either through bare soil or through a polyethylene tarpaulin represented by a diffusive transfer coefficient h.

4. Fumigant is initially present in a narrow band of concentration at a depth L in the soil. This is an idealization of the dispersed initial distribution resulting from the pressure-driven and density-driven early stages.

5. A barrier to gaseous diffusion is present at $z = P$. This depth may be finite or infinite.

6. The fumigant partitions linearly and instantaneously into the dissolved and sorbed phases.

7. The fumigant undergoes first-order degradation in the soil, described by a degradation rate coefficient μ that is constant and independent of location.

Under these assumptions, we can write the transport equations in layers 1 and 2 as follows:

$$R_1\frac{\partial C_1}{\partial t} + \mu R_1 C_1 = D_1\frac{\partial^2 C_1}{\partial z^2}; \qquad 0 < z < H \tag{1}$$

$$R_2\frac{\partial C_2}{\partial t} + \mu R_2 C_2 = D_2\frac{\partial^2 C_2}{\partial z^2}; \qquad H < z < P \tag{2}$$

where C_i, i=1,2 are the gas-phase concentrations in the respective layers, D_i are the soil-gas diffusion coefficients, described by

$$D_i = \xi(a)D_g^a = (\frac{a_i^2}{\phi^{2/3}})D_g^a \qquad i = 1,2 \tag{3}$$

where ξ is the tortuosity (7), a is soil air content, and D_g^a is the binary diffusion coefficient of the fumigant in air. R_i are the gas-phase partition coefficients, described by

$$R_i = \frac{a_i K_H + \theta_i + \rho_b K_d}{K_H} \qquad i = 1,2 \tag{4}$$

where ρ_b is soil bulk density, K_H is Henry's constant and K_d is the soil sorption distribution coefficient for the chemical.

The initial, surface, and lower boundary conditions are expressed as follows:

$$C_1(z,0) = 0 \tag{5}$$

$$C_2(z,0) = \frac{M}{R_2}\delta(z - L) \tag{6}$$

$$-D_1\left\{\frac{\partial C_1}{\partial z}\right\}_{z=0} + hC_1(0,t) = 0 \tag{7}$$

$$-D_2\left\{\frac{\partial C_2}{\partial z}\right\}_{z=P} = 0 \tag{8}$$

where M is the initial mass injected at $z = L$, $\delta(z - L)$ is the Dirac-delta function representing a narrow initial pulse of chemical at $z = L$, and h is the transfer coefficient through the polyethylene tarp covering the surface. For the case of a bare surface, $h \to \infty$.

The two regions are linked by conditions of concentration and flux continuity at $z = H$

$$C_1(H,t) = C_2(H,t) \tag{9}$$

$$- D_1 \left\{ \frac{\partial C_1}{\partial z} \right\}_{z=H} = -D_2 \left\{ \frac{\partial C_2}{\partial z} \right\}_{z=H} \tag{10}$$

The solutions to the above equations for the soil concentration, the surface volatilization flux, the concentration-time index, and the cumulative volatilization losses are given in Appendix A.

Results

Homogeneous Soil Figure 1 shows calculated volatilization fluxes as a function of time for various soil conditions following methyl bromide injection. Standard conditions for each calculation unless varied are air content $a = 0.4$, injection depth $L = 20$ cm, barrier depth $P = \infty$, tarpaulin transfer coefficient $h = 2.4$ cm h^{-1}. The patterns are predictable, but some features are worth pointing out.

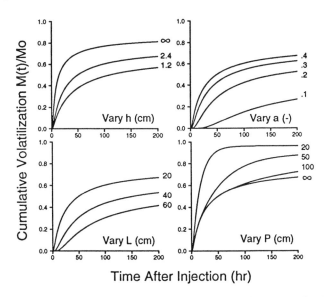

Time After Injection (hr)

Figure 1: Cumulative volatilization losses to the atmosphere in soil with uniform water content as a function of the variables indicated on the figure caption.

The upper left figure shows the effect of no tarp (h=∞), compared with 1- and 2-mil thick tarps (h=2.4, 1.2 cm h^{-1}, respectively (5)). Because the tarp is permeable to methyl bromide, losses are not substantially reduced when injection is shallow, although they are delayed somewhat. Injection depth (lower left figure) influences losses, but only reduces cumulative volatilization from 73 to 52% when lowered from 20 to 60 cm. Air content variations (upper right figure) in the range of 0.2 to 0.4 have a surprisingly small effect on volatilization (from 73 to 61%) when injection depth is shallow, because the residence time in the soil is still small compared to the decay time even when the soil is quite wet. There is a substantial reduction in volatilization (to 40%) when the air content is lowered further to 0.1.

Barrier depth has a pronounced effect on volatilization losses, because a barrier prevents downward diffusion. When the barrier is at 20 cm, essentially all (96%) of the injected methyl bromide is lost.

Irrigated Soil When soil is irrigated after methyl bromide injection but before tarping, the effect is to produce an additional barrier to upward movement. Since the tarp effectively prevents water loss from the soil, the volume of water irrigated acts in some respects like a layer of higher diffusion resistance. This is modeled in our calculations as a two-layer soil.

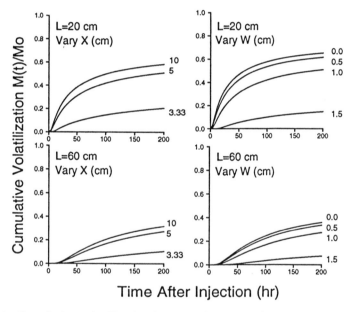

Time After Injection (hr)

Figure 2: Cumulative volatilization losses to the atmosphere in soil with nonuniform water content following irrigation as a function of the variables indicated on the figure caption.

Figure 2 shows the effect of water volume W and depth of infiltration X of a given quantity of water on volatilization from soil following shallow and deep injection of methyl bromide. As shown in the right-hand figures, water volume has only a modest effect for a 5-cm penetration until the quantity of water added is high (1.5 cm). This is similar to the experimental result observed by Jin and Jury (5), who lowered volatilization loss from essentially 100% to a few percent when 1.6 cm of water was added to their volatilization column. For a given amount of water added, the depth of penetration of the front is important, as shown in the left-hand figures. A 1 cm irrigation in the shallow injection system results in about 66% cumulative loss when this volume of irrigation water penetrates to 10 cm, but only 29% loss if the same irrigation volume only penetrates 3.3 cm. Clearly, nonuniform water content influences volatilization loss proportionately more than lowering the entire water content uniformly. The effect of having a high diffusion resistance near the surface is to promote downward diffusion, where the gas degrades before migrating back to the surface

Concentration-Time Index   Reducing volatilization loss of a soil fumigant will be of no benefit if the fumigant's efficacy at weed, insect, or pathogen control is compromised in the process. Current commercial fumigation guidelines are based largely on the results of early studies and require keeping the subsurface soil "as dry as possible". However, it has been frequently observed that many soil pathogens and fungi are more susceptible to control by MeBr in wetter soils, because partition into the water phase is required for MeBr to exert its biocidal action (8-10). The optimum soil water content is also determined by the types of disease or insects to be treated and the depth where they reside. Most of the nematodes, fungi and pathogens are located primarily at shallower depths, so higher water contents can be used. The optimum moisture level for a specific soil to be treated with MeBr is determined to a great extent by soil texture. It is difficult to achieve effective fumigation in heavy textured soils even at low water contents. However, much higher soil water contents than those recommended by current fumigation guidelines can be used for coarser-textured soils.

Different MeBr exposure levels are required to control different pests, with the highest amounts required for controlling nematodes, fungi, and pathogenic bacteria (6). The cumulative dosage or concentration-time index (CT), which is the product of the average concentration and the exposure time, has been used as a criterion to evaluate the effectiveness of a fumigation event. Thus, subjecting the target organism to a lower concentration over a longer exposure time, which would result if the soil surface was restricted by a water barrier, could achieve the same CT index as having short exposure to high concentrations, the strategy that is currently being used in conventional fumigation operations.

We calculated the concentration-time index as a function of position and time to compare the efficacy of various fumigant management alternatives. Figure 3 shows the CT at the end of 3 weeks for tarped and bare homogeneous soil. It is clear that the main difference between these two methods is near the surface, where the bare-soil treatment has significantly lower CT. It appears that the principal benefit of tarping may be in allowing soil concentrations near the surface to build up. This is expected, because it has been found that although the polyethylene tarp does not prevent MeBr emission, it does delay its release into the atmosphere (5). Figure 4 compares CT at 3 weeks for a 1-cm irrigation penetrating to different depths. Except very near the surface, the CT is higher in all cases for lesser degrees of penetration.

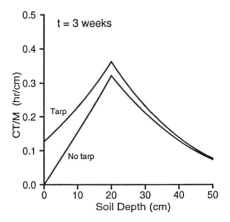

Figure 3: Concentration time index accumulating over three weeks in tarped and untarped soil of uniform air content $a = 0.35$ and injection depth $L = 20$ cm.

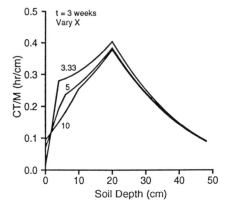

Figure 4: Concentration time index accumulating over three weeks in tarped soil of air content $a = 0.35$ and injection depth $L = 20$ cm irrigated with 1 cm of water as a function of the depth of penetration X of the water front.

The buildup of CT over time is shown in Figure 5 for homogeneous and irrigated soil. In both cases, much of the CT is accumulated during the first week, with lesser benefits thereafter. The principal difference in CT between homogeneous and irrigated soil is in the wetter soil zone and immediately below it. Although it appears that the wetter soil in the irrigated zone may be excluding fumigant significantly, the CT indices are not affected much. Figure 6 shows the CT at 2 cm as a function of time for four different management options. As shown, the two irrigation treatments have comparable CT to tarped soil, whereas the bare soil treatment's CT is significantly smaller.

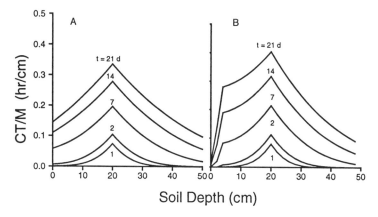

Soil Depth (cm)

Figure 5: Concentration time index at various times in tarped soil of (A) uniform air content $a = 0.35$ and (B) with $W = 1$ cm irrigation penetrating to $X = 5$ cm with injection depth $L = 20$ cm.

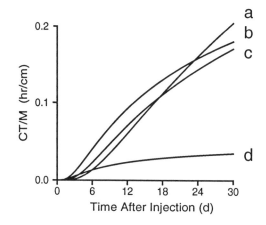

Time After Injection (d)

Figure 6: Concentration time index as a function of time at z=2 cm depth in soil of air content $a = 0.35$ and injection depth $L = 20$ cm. (a) 1 cm irrigation under tarped soil entering to depth $X = 3.33$ cm; (b) tarped soil with no irrigation; (c) 1 cm irrigation under tarped soil entering to depth $X = 5.0$ cm; (d) untarped soil with no irrigation.

Discussion

Field Observations The field studies mentioned in the introduction had widely different observed rates of MeBr loss to the atmosphere. The soil in which Yagi et al. (1) performed their study differs from the others in that it has a zone of high water content and high bulk density at about 30 cm that could have acted as a barrier to downward movement (2). This might explain the high loss observed in their first study (see lower right of Fig 1). In the second study by this group, the water content of the upper zone was higher and the injection depth lower, both of which are consistent with the results in Figure 1.

The studies by Majewski et al. (3) clearly occurred in wetter than normal soil because of the time of year (November), after some rain had fallen in the area. Under these conditions, the tarp would restrict water evaporation and prevent the soil from drying out, which might account for the differences observed between the bare and tarped soil. The low rate they found under tarped soil (32%) could simply be a consequence of the higher resistance the soil offers when wet.

The study by Yates et al. (4) under tarped soil occurred under dry summer conditions with no barrier to downward movement. It is reasonable therefore that the loss they observed should be intermediate between the summer experiment of Yagi et al. (1) with a barrier present, and the wetter late fall study of Majewski et al. (3).

Effect of Infiltration Our model calculations showed that a water layer near the surface was a substantial deterrent to volatilization loss. This use of supplemental irrigation water to prevent MeBr loss was motivated by the experiment of Jin and Jury (5), where a modest 1.6 cm addition of water lowered volatilization to insignificant levels. Yet, the model calculations also showed that uniformly increasing water content was far less effective than a shallow water layer at restricting volatilization, so that redistribution following infiltration might (in a coarse soil) quickly erase nonuniformities in the profile.

We believe that this will not happen under field conditions, for rather subtle reasons. The tarp is a barrier to water loss, and we have observed that the underside tends to act as a condensation surface (Figure 7), resulting in droplets being redeposited on the surface. Further, diurnal heat fluctuations tend to bring water vapor upward to the surface at night. These two effects in concert will counteract the normal tendency of gravity to smooth out a profile following the end of irrigation.

We tested this hypothesis in a laboratory column in which a 1-cm irrigation was added to soil of uniform water content and then sealed with a polyethylene tarp. After 4 weeks of exposure to diurnal heat fluctuations we sampled the soil column and found evidence of water accumulation at the surface (Figure 8).

The water addition may also have its greatest value at early times, when much of the observed MeBr loss occurs. The early time behavior of MeBr is influenced by pressure-driven expansion as well as diffusion, and the added resistance of the water layer could have a significant effect on volatilization. We are currently field-testing this hypothesis.

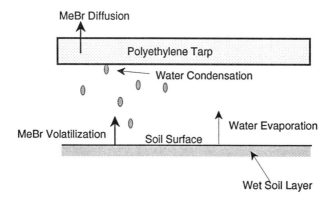

Figure 7: Schematic diagram of the water circulation process under a tarpaulin.

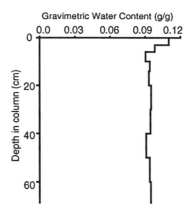

Figure 8: Gravimetric water content in a soil column four weeks after receiving a 1 cm irrigation and then being sealed under a tarpaulin and exposed to diurnal heat cycling.

Appendix A

The general solution in Laplace transform space to equations 1,2,5, and 6 is given by

$$\widehat{C}_1(z) = A_1 \exp(q_1 z) + B_1 \exp(-q_1 z) \tag{11}$$

$$\widehat{C}_2(z) = A_2 \exp(q_2 z) + B_2 \exp(-q_2 z)$$

$$+ \ U(z-L)\frac{M}{2q_2D_2}[\exp(q_2(L-z)) - \exp(-q_2(L-z))] \qquad (12)$$

where A_1, A_2, B_1, B_2 are constants, $q_i = \sqrt{R_i(s+\mu)/D_i}; i = 1, 2$, U is the Heaviside unit function, and s is the Laplace transform variable. By applying Eq 7, we can rewrite (11) as

$$\hat{C}_1(z) = A_1[\exp(q_1z) + \Omega \exp(-q_1z)] \qquad (13)$$

where

$$\Omega = \frac{q_1D_1 - h}{q_1D_1 + h} \qquad (14)$$

Similarly, we can use (8) to rewrite (12) as

$$\begin{aligned} \hat{C}_2(z) &= B_2[\exp(q_2(z-2P)) + \exp(-q_2z)] \\ &+ \frac{M}{2q_2D_2}U(z-L)[\exp(q_2(L-z)) - \exp(-q_2(L-z))] \\ &+ \frac{M}{2q_2D_2}[\exp(q_2(z-L)) + \exp(q_2(z+L-2P))] \end{aligned} \qquad (15)$$

We then apply the continuity conditions (9)-(10) at $z = H$ to solve for A_1, B_2. The final equations for the concentration may be written as

$$\hat{C}_1(z) = M\frac{\cosh[q_2(P-L)]\{\sinh(q_1z) + \sigma\cosh(q_1z)\}}{q_1D_1\{\alpha_1\cosh[q_2(P-H)] + \gamma\alpha_2\sinh[q_2(P-H)]\}} \qquad (16)$$

$$\begin{aligned} \hat{C}_2(z) &= M\frac{\cosh[q_2(P-L)]\{\alpha_1\sinh[q_2(z-H)] + \gamma\alpha_2\cosh[q_2(z-H)]\}}{q_2D_2\{\alpha_1\cosh[q_2(P-H)] + \gamma\alpha_2\sinh[q_2(P-H)]\}} \\ &+ \ MU(z-L)\frac{\sinh[q_2(L-z)]}{q_2D_2} \end{aligned} \qquad (17)$$

where

$$\alpha_1 = \cosh[q_1H] + \sigma\sinh[q_1H] \qquad (18)$$

$$\alpha_2 = \sinh[q_1H] + \sigma\cosh[q_1H] \qquad (19)$$

$$q_i = \sqrt{\frac{R_i(s+\mu)}{D_i}}; \qquad i = 1, 2 \qquad (20)$$

$$\sigma = \sqrt{\frac{R_1D_1}{h^2}} \qquad (21)$$

$$\gamma = \sqrt{\frac{R_2D_2}{R_1D_1}} \qquad (22)$$

Equations 16 and 17 may be used to calculate other quantities of interest in transform space, such as the volatilization flux

$$\hat{J} = h\hat{C}_1(0) \tag{23}$$

or the concentration-time index

$$\widehat{CT}(z) = \frac{\hat{C}_1(z)}{s} \tag{24}$$

Moreover, the cumulative volatilization loss from the surface

$$V_c(\infty) = \int_0^\infty J(t')dt' \tag{25}$$

may be calculated from the transform expressions as follows

$$V_c(\infty) = \hat{J}_{s\to 0} \tag{26}$$

Special cases of the general solution given in the equations above may be obtained by taking the following limits:

- No tarpaulin $\sigma \to 0$ $(h \to \infty)$
- Homogeneous soil without irrigation $q_1 = q_2$; $\quad \gamma = 1$
- No barrier $P \to \infty$

In all cases, the Laplace transform solutions were inverted by the method of Talbot (11). A fortran program of this routine is given in Jury and Roth (12).

Literature Cited

1. Yagi, K.; Williams, J.;Wang, N.-Y.; Cicerone, R. J. *Proc. Natl. Acad. Sci.* 1993 90,8420-8423.
2. Yagi, K.; Williams, J.;Wang, N.-Y.; Cicerone, R. J. *Science* 1995 267,1979-1981.
3. Majewski, M. S.; McChesney, M. M. ;Woodrow, J. E.; Prueger, J. H.; Seiber, J. N. *J. Environ. Qual.* 1995 24,742-752.
4. Yates, S. R.; Ernst, F. F.; Gan, J.; Gao, F.; Yates, M. V. 1996. *J. Environ. Qual. 25, 192-202*
5. Jin, Y.; Jury,W. A. *J. Environ. Qual.* 1995, 24,1002-1009.
6. Chakrabarti, B.; Bell, C. H. *Chemistry & Industry,* 1993, 24,992-995.
7. Jin, Y.; Streck, T.; Jury,W. A. *J. Contaminant Hydrol.* 1994, 17,111-127.
8. Kolbezen, M. J.; Munnecke, D. E.; Wilbur, W. D.; Stolzy, L. H.; Abu-El-Haj, F. J. ; Szuszkiewicz., T. E. *Hilgardia* 1974, 42,465-492.
9. Krikun, J.;Netzer, , D.;Sofer, M. *Agro-Ecosystems* 1974, 1,117-122.
10. Munnecke, D. E.;Moore, B. J.; Abu-El-Haj, F. J. *Phytopathology*1971 61,194-197.
11. Talbot, A. *Inst. Math Appl.* 1979 23,97-120.
12. Jury, W. A.; Roth, K. *Transfer Functions and Solute Movement through Soil: Theory and Applications. Birkhäuser Publ.: Basel, 1990.*

Chapter 11

Emissions of Methyl Bromide from Agricultural Fields: Rate Estimates and Methods of Reduction

S. R. Yates[1], Jianying Gan[1], F. F. Ernst[1], D. Wang[1], and Marylynn V. Yates[2]

[1]Soil Physics and Pesticide Research Unit, U.S. Salinity Laboratory, Agricultural Research Service, U.S. Department of Agriculture, 450 West Big Springs Road, Riverside, CA 92507
[2]Department of Soil and Environmental Sciences, University of California, 2208 Geology, Riverside, CA 92521

Methyl bromide, a soil fumigant, is under intense scrutiny due to evidence which suggests that it damages the stratospheric ozone layer. Because of this, methyl bromide is scheduled for phase-out by 2001. The National Agricultural Pesticide Impact Assessment Program has determined that there will be substantial adverse economic impacts on the agricultural community if the use of methyl bromide is restricted. This has prompted numerous scientists to: study the environmental fate and transport of methyl bromide; search for replacement chemicals and/or nonchemical alternatives; and develop new methodology which improves containment of methyl bromide (or any alternative fumigant) to the treatment zone, while maintaining adequate pest control. This paper reports on several recent experiments to measure of methyl bromide emissions from agricultural operations. Information is also provided on the processes and mechanisms which must be fully understood if reliable methods for reducing atmospheric emissions are to be obtained, without a reduction in pest control.

For decades, methyl bromide (bromomethane, MeBr) has been used for the control of nematodes, weeds and fungi. Recently, it has come under scrutiny as a chemical which depletes stratospheric ozone. Under the provisions of the U.S. Clean Air Act, which calls for the discontinuation of compounds which deplete ozone, MeBr is scheduled for phase-out by the year 2001. The USDA National Agricultural Pesticide Impact Assessment Program *(1)* conducted an assessment of the economic impact of eliminating MeBr use and determined that there will be a substantial adverse impact on the agricultural community. These effects will be most strongly felt in two states, California and Florida, which are the primary users of MeBr. It has estimated *(1)* that a MeBr phase-out in soil fumigation will cause $1.5 billion dollars in annual lost production in the United States. This estimate is conservative, however, since it

0097–6156/96/0652–0116$15.00/0
© 1996 American Chemical Society

ignores post-harvest, non-quarantine uses and quarantine treatments of imports and other future economic aspects such as lost jobs, markets, etc. In terms of specific commodities, major crop losses would occur with tomatoes ($350 M), ornamentals ($170M), tobacco ($130M), peppers ($130M), strawberries ($110M) and forest seedlings ($35M).

Over the past decade, concern has increased that halogenated gases emitted into the atmosphere are destroying the stratospheric ozone layer. The Ozone Assessment Synthesis Panel of the United Nations Environmental Programme (UNEP) states that the hole in the Antarctic ozone layer is due primarily to increases in chlorine- and bromine-containing chemicals in the atmosphere. Although 90 to 95% of the ozone loss is thought to be from chlorinated compounds *(2)*, attention has been focused more recently on MeBr because bromine is believed to be 40 times more efficient than chlorine in breaking down ozone on a per atom basis *(3)*. Although the largest effects from ozone-depleting gases have been observed in the southern hemisphere, there are indications that atmospheric ozone is also decreasing in the northern hemisphere.

To complicate matters, there is a great deal of uncertainty in the estimates of global sources of bromine. For example, it has been estimated that natural bromide-gas production by marine plankton in the oceans contributes 50-80% of the global burden. Agricultural fumigation, however, represents approximately 15-35% *(4-7)* and recent figures indicate that biomass burning may contribute up to 30% *(8)*. The oceans may act as a net sink, rather than a source *(9)* of bromine-gases; and the deposition onto soil and subsequent microbial degradation may be another important pathway for removing MeBr from the atmosphere *(10)*. Also, although agricultural emissions represent a significant fraction (i.e., 15 to 35%), even if MeBr is no longer used in agriculture, large amounts of bromine-gases will continue to exist in the atmosphere and, therefore, must be considered a natural condition. Even so, it is desirable to develop improved methods for reducing agricultural MeBr (and alternatives) emissions to the atmosphere so that anthropogenic contributions are minimized.

In this paper, recent measurements of MeBr emissions under field conditions are summarized, and a field study in which three independent methods were used to obtain the emissions rate is described. Based on recent field and laboratory studies and published information, approaches for reducing MeBr emissions are also discussed.

Measured Emissions Rates

There have been several recent experiments conducted to obtain information on MeBr emissions from typical agricultural operations. These studies used various methods for estimating the emission rate and include: an increase in soil Br^- concentration as a result of MeBr degradation *(11)*, atmospheric flux method *(12,13)* and enclosed flux chamber method *(14-16)*. Each method has advantages and disadvantages which can make the interpretation of the experimental results somewhat difficult. The Br^--appearance method assumes that the difference between the MeBr mass applied and mass degraded (i.e., Br^- produced) was released into the atmosphere. Therefore, measuring Br^- in the soil provides a method for estimating the total atmospheric emission. An advantage of this method is the ease of analyzing the

Br⁻ content of soils. A disadvantage is the large number of soil samples necessary to obtain an accurate field-scale estimate of degradation at all depths (11). Also, no information about the dynamics of MeBr emissions can be obtained using this method. Atmospheric flux methods are fairly complex, require numerous measurements of MeBr concentration and other meteorological parameters and may require assumptions concerning the behavior of the atmosphere. Advantages are that the methods are well tested, they provide a field-scale average total emission rate and they provide information on the dynamics of the volatilization process. The flux chamber method (17-19) is one of the simplest methods for measuring pesticide flux, but it suffers from several disadvantages. The method measures the flux over a small area which can cause the estimated flux rate to be highly variable, the flux estimates are sensitive to the placement of the chambers relative to the position of MeBr injection (i.e., distance to the source), and the presence of chamber can affect the area sampled (especially the local temperature and relative humidity). These can have a tremendous effect on experimental uncertainty.

Yagi et al. (14) conducted an experiment to measure the MeBr emission from a southern California field using passive flux chambers. MeBr was applied at approximately 25 cm depth and the soil surface was covered with polyethylene plastic. The authors estimated that 87% of the total MeBr applied to the field escaped into the atmosphere. This is the highest reported estimate for MeBr emissions when the compound was injected at shallow depth and the field was covered with plastic. The high emission rate may have been due, in part, to the high bulk density of the soil and the presence of a wetter soil layer at 60 cm depth. The authors indicated that this value was higher than expected given other estimates based on mathematical models (20,15), but was similar in magnitude to the losses observed in glass-house studies (21). To verify these results the authors returned to the field to collect Br⁻ information to provide a rudimentary mass balance estimate (15). In addition, these investigators conducted a second experiment using the same procedures as their first experiment and found that only 34% of the applied MeBr escaped to the atmosphere. This value is 61% lower than the result of their first experiment. This sort of variability is not unexpected for several reasons: 1) only 10-15 samples of the volatilization rate were obtained during each 7-day experiment, generally at the high point during the day; 2) only a few soil samples were taken to measure Br⁻ concentrations and soil Br⁻ concentration has been shown to be highly variable (11,22); 3) the soil Br⁻ concentration after fumigation was measured to only 90 cm; and 4) for the first experiment, initial Br⁻ concentration was available only at depth of 3.0 cm and was extrapolated downward. An additional source of variability may be the internal chamber temperature. Yates et al. (16) demonstrated that chambers can produce erroneously high volatilization rates if their presence on the tarp causes an increased internal chamber temperature relative to the outside environment. Yagi et al. (14,15) did not correct their volatilization rates for this effect.

In a study conducted in a strawberry field, Majewski et al. (12) found that 32% of the applied MeBr was emitted into the atmosphere during the first 6 days following application. This value is approximately the same as that from the second study of Yagi et al. (15). The MeBr application rate for this experiment was 392 kg/ha and the flux density was measured using the aerodynamic method *(23)*. The reported total loss fell into the 30-60% range noted in the Montreal protocol *(20)*, but a mass balance was not conducted. More information on this experiment is given in this proceedings.

An Experiment with Three Independent Measures of Total Emissions

An experiment *(11,13,16)* was conducted at the University of California's Moreno Valley Field Station on a 4-ha field between August 26, 1993 and September 13, 1993. The soil in this field is a Greenfield sandy loam. MeBr (applied as 99.5% MeBr (CH_3Br) and 0.5% chloropicrin (CCl_3NO_2)) was applied at a shallow depth of 25 cm, at a rate of 240 kg/ha, and the field was covered with 1 mil polyethylene plastic. Three independent methods were used to give estimates of the MeBr emission rate and total loss.

Estimating Total Loss from Br^- Appearance.   To estimate the total MeBr mass converted to Br^-, numerous soil cores were taken to a maximum depth of 7 m. Four soil cores were taken in the center of the field to a depth of 2 m and one to 3 m prior to applying MeBr to provide background concentrations. The Br^- concentrations were measured using an ion selective electrode connected to an Accumet 25 pH meter (Fisher Scientific Co.) at 0.3-m depth increments. After the experiment, 25 cores to a depth of 5 m and 5 cores to 7 m were taken randomly in the field. These cores were sectioned at 0.1-m intervals from the surface to a depth of 1.0 m and at 0.2-m intervals from 1 to 7 m. Comparing the pre- and post-treatment Br^- concentrations, it was determined that additional background concentrations were needed to reduce spatial variability and improve the accuracy of the estimate of the background Br^- concentration. Therefore, 30 additional soil cores to 7 m were obtained in the field adjacent to the experimental site which had the same soil type, cropping and irrigation history but was never fumigated.

Figure 1 shows the background Br^- [mg/kg] concentration on a dry soil weight basis taken prior to application (open squares). The samples taken from the adjacent field are shown as closed diamonds. Also shown on these curves is a bar which indicates an average standard deviation for the curve calculated by averaging the standard deviations at each depth and, therefore, is the same at every point. The Br^- concentration 36 days after application is shown as open circles. An estimate of the total MeBr lost to the atmosphere can be obtained from the difference between the initial and final curves and converting from Br^- mass to MeBr mass.

Using the information in Figure 1, 325($\pm$ 164) kg or 39% ($\pm$ 19%) of the applied MeBr was degraded to Br^-. Since the MeBr mass remaining in the field was estimated to be less than 0.05% at the time of sampling, the total loss from volatilization is estimated to be approximately 518 ($\pm$164) kg. The spatial and measurement variability introduces uncertainty into the Br^- mass calculation as shown by the standard error of $\pm$164 kg. Uncertainty in the measured Br^- directly affects the certainty in the estimate of MeBr volatilized from the field, producing 61% $\pm$ 19%.

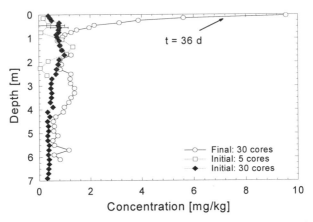

Figure 1. Bromide ion concentration as a function of depth in the field.

Field scale variability must be considered when obtaining the average Br⁻
concentration in the field. This is especially important when this information is used in
estimating the total MeBr lost from the field. This can be illustrated as follows. If the
estimate of the mass is obtained using the 5 background soil cores to a depth of 2 and 3
m (i.e., 45 samples), the total loss is estimated to be 298 kg or 35.3%. If in addition,
the samples for depths below 3 m taken from the 30 cores located in the adjacent field
are used to extend the initial distribution below 3 m (i.e., 500 samples), the total loss is
estimated to be 435 kg or 48.4%. When only the 30 soil cores are used and all depths
considered (i.e., 1100 samples), the mass loss is estimated to be 518 kg or 61%. This
demonstrates that numerous deep soil cores are needed to adequately estimate MeBr
degradation in soil and that the 5 soil cores from the field interior happen not to
produce an accurate field-scale average of the initial Br⁻ concentration. The estimate
of the field-average Br⁻ mass which makes use of the most samples has the highest
probability of being the most accurate.

Estimating Emissions from Atmospheric Flux Methods. To collect air samples
above the fumigated field which could be analyzed for MeBr, a sampling mast was
constructed *(24)* and placed in the center of the field. The mast held coconut-based
charcoal sampling tubes at heights of 0.1, 0.2, 0.5, 0.8, 1.2, and 1.6 m above the field
surface. A vacuum system was used to draw air (at 100 mL/min) through the sampling
tubes to extract the MeBr gas. The duration of the sampling intervals was either 2 or 4
hours. The atmospheric concentration and weather conditions were continuously
monitored 24 hours a day until the air concentrations dropped below detectable limits.
The method used to analyze the charcoal sample tubes and the details of the error
analysis resulting from sample handling are given by Gan et al. *(25,26)*.

The aerodynamic *(23)*, theoretical profile shape *(27,28)* and integrated
horizontal flux *(29)* methods were used to estimate the total MeBr emission. Since
these flux-estimation techniques use the same gas concentration data, they do not

represent completely independent flux estimates. However, if the three methods produce similar emission rates, this would be supportive evidence of valid experimental procedures.

Figure 2a shows the flux density during the first 7 days of the experiment and Figure 2b reports the cumulative mass lost. The solid line was obtained from the aerodynamic method.

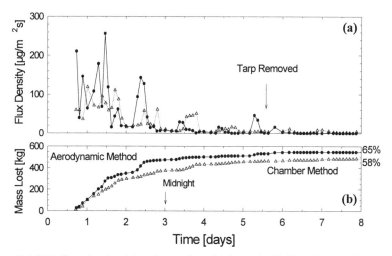

Figure 2. MeBr flux density (a) and mass lost (b) from the field during the first seven days of the experiment.

The highest flux density occurred at the beginning of the experiment when nearly 36% of the applied MeBr mass was lost during the first 24 hours after application. The highest flux rates occurred during the late morning and early afternoon when temperatures were highest and the atmosphere was unstable. Cooler temperatures, light winds and neutral to stable atmospheric conditions were present at night; generally reducing the flux. Using the aerodynamic method, the total emission was estimated to be 62% (±11%) to 67% (±6%) of the mass applied to the field. For the theoretical profile shape and integrated horizontal flux methods, respectively, the estimates were 61% ± 3% (of applied) and 70% ± 3% (of applied). A mass balance was calculated for each method used to estimate the flux (Table 1). The average mass recovery using all the flux methods was 867 kg (±83 kg), which was 103% (±10%) of the applied mass (i.e., 843 kg). The range in the mass balance percent (i.e., percent of applied mass that is measured) was from 97% to 108%. The averaged mass balance percent for the discrete aerodynamic method, which involved using the measured data directly, was approximately 101%.

Table 1. Total Amount of MeBr Volatilized During the Experiment and Mass
          Balance

Flux Method Used	Mass Lost [kg]	Percent Lost [%]	Mass Balance[†] [%]
[‡]Aerodynamic, (discrete)	525 (± 91)[§]	62.2	101
[‡]Aerodynamic, (profile)	568 (± 47)	67.3	106
[‡]Theoretical Profile Shape	506 (± 29)	60.1	99
[‡]Integrated Horizontal Flux	588 (± 21)	69.8	108
[¶]Flux Chamber, (corrected)	464 (± 170)	58.8	97

† Mass Applied 843 kg; Mass Remaining 0.26 kg; Mass Degraded 325 kg
‡ Data from *(13)*
¶ Data from *(16)*
§ Values in parentheses are standard deviations

The fraction of the applied mass lost from this experiment is approximately
double the value reported by Majewski et al. *(12)* who estimated the total loss to be
approximately 32%. This is probably due to differences in the climatic and soil
conditions between the Monterey region and Moreno Valley. Lower temperatures in
Monterey would cause a reduction in the diffusion through polyethylene plastic material
(30) and increase the residence time in the soil. This would facilitate greater MeBr
degradation in the soil and reduce the total loss to the atmosphere. The range for total
emissions described herein also differs from the results of Yagi et al. *(14,15)* who
reported values of approximately 87% and 34%, respectively, for experiments with a
similar MeBr application methodology.

Estimating Emissions using Flux Chambers. An independent estimate of flux was
obtained using three flow-through chambers *(31,18)*. The MeBr volatilization rate and
cumulative mass lost from the field was obtained by integrating the chamber flux
density data shown in Figure 2 (dashed line) over the entire field and course of the
experiment. During the first 24 hours after application, approximately 365 kg or 45%
of the applied mass volatilized from the field. During the next 24 hours, an additional
202 kg or 25% was lost to the atmosphere.
          The total mass emitted from the field was estimated to be 811 kg, which is
about 96% of the 843 kg that was applied to the field. The mass balance calculated
using this data was 135%, which was not consistent with the estimates derived from

the soil Br⁻ data and atmospheric flux methods. This discrepancy prompted an investigation of the flux chamber data.

The air temperature inside the chambers used in the experiment was found to be much higher than the air temperature outside and was highly correlated with the diurnal variation in incoming solar radiation. Figure 3 shows the MeBr flux density through polyethylene film in response to changes in the ambient temperature. The plastic used during this experiment is shown as open boxes (other data from (*30*)). Using this information, a method was developed to correct the chamber flux density data for enhanced flux caused by increases in the temperature inside the chamber *(16)*.

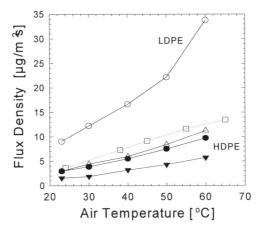

Figure 3. Temperature dependence of the flux density through polyethylene. Higher flux density is equivalent to higher emissions.

After correcting for temperature, the total mass emitted from the field was estimated to be about 496 (± 175) kg as opposed to 811 (± 303) kg. The loss represents about 59% of the total applied mass, which more closely follows the results from the other estimates. During the first 24 hours after application, approximately 227 kg (27% of applied) of MeBr was lost, which is 46% of the total emissions. During the next 24 hours, an additional 117 kg (14% of applied) was lost. The corrected total mass lost is about 3 to 10% lower than the estimates from the other methods. A mass balance of 97% was obtained for the corrected measurements (Table 1).

Factors Important in Reducing Emissions

There are many soil-chemical processes which affect the fate and transport of fumigants, including MeBr. Generally, three factors must be controlled to reduce emissions while maintaining adequate efficacy: containment, degradation and soil-gas concentration (i.e., effective dosage). Unless each of these factors is controlled, unacceptable emissions will likely result. For example, in the absence of degradation, perfect containment alone will not produce lower emissions unless the field remains covered indefinitely. A balance must be achieved with adequate containment together with sufficient degradation to reduce the amount of MeBr in the soil prior to removing the plastic cover, all of this, while maintaining adequate soil concentration levels to control pests.

Laboratory Experiments. Soil columns were used to determine how injection depth, use of plastic covers, soil water content, bulk density and soil organic matter affects the total MeBr loss from soils. The columns were 60 cm in length and have a closed bottom which restricts downward diffusion of MeBr. This restriction causes the volatilization rates to be overestimated when compared to an infinite-length column which is analogous to the field. A diffusion model was used to correct the emission rates so that they relate more closely to field situations. The corrected results are used in the discussion below. In brief, four steps were involved. 1) Experiments were conducted to obtain the emission rate and the soil-gas concentration at different times for the selected management factors. 2) Multiple sets of the measured MeBr concentrations in soil air were used in a gas-diffusion transport model to obtain the model parameters under the experimental conditions (e.g., when an impermeable barrier occurs at 60 cm). 3) These parameters were used in a similar model to estimate MeBr volatilization rates for columns without a barrier, which is analogous to the field. 4) The results from the two models were used to obtain the ratio: (simulated total loss without barrier)/(simulated total loss with barrier) and the measured laboratory values were multiplied by this ratio to give an estimate of the volatilization rate in a field soil experiencing similar conditions.

Containment. Containment is necessary to hold the gas at the treatment location and provide sufficient time for pest control. Without adequate containment, significant fractions of applied MeBr will be lost to the atmosphere. The need for containment is due to MeBr's high mobility as a result of its high vapor pressure (approximately 1420 mmHg at 20°C) and low boiling point (3.56°C). Because of these properties, a large fraction of MeBr exists in the vapor phase at temperatures and pressures that normally occur in the field. Since the gas-phase diffusion coefficient is nearly 10,000 times greater than in the liquid phase *(32)*, pesticides which have a large vapor pressure easily move through soil *(33-35)*. Factors that affect containment include: use of plastic, the properties of plastic, injection depth, soil bulk density, soil water content, soil cracking, and other mechanisms which promote or retard movement. For example, shortly after injection, pressure-driven flow may dominate MeBr movement in response to phase-change expansion and the initially high gradients near the injection point. This can cause MeBr to quickly move to the soil surface where it can escape into the atmosphere. Other processes may also be important in moving fumigants through the root zone. For example, changes in barometric pressure *(36)*, pressure effects caused from wind at the surface and density sinking *(33)* all may induce a mass flow. While it may be possible to take advantage of many soil factors to aid in containing MeBr, the inherent spatial variability of soils make it difficult to ensure emissions control for every situation.

Plastic Films. Probably the most common and the most predictable method to improve containment and reduce the amount of MeBr leaving the treated soil is the use of plastic films. Covering the field with plastic can reduce the amount of MeBr volatilized by inhibiting transport from the soil into the atmosphere. Advantages of using films are that the properties and condition of the film is known in advance and films are more uniform in space and time compared to soil. Therefore, there may be a higher certainty of effective containment when films are used compared to soil-water based methods. Also, the level of containment can be controlled by altering the plastic material used. For example, new plastics are available which are highly impermeable to MeBr diffusion (Table 2).

Table 2. Flux Density $[mg/m^2/h]^a$ of fumigants through 1.4-mil high-density polyethylene film and 1-mil Hytibar® Film[b]

Fumigant	Flux Density[a] Through 1.4-mil Polyethylene $[mg/m^2/h]$	Flux Density Through 1-mil Hytibar[b] $[mg/m^2/h]$	Material	Methyl Bromide Flux Density $[mg/m^2/h]$
Methyl Bromide	7.4	0.1	2-mil silver mylar	1.4
Methyl Iodide	8.9	0.06	2-mil mylar	2.2
(E) 1,3-D	87	0.2	5-mil mylar	2.1
(Z) 1,3-D	62	0.2	1-mil polyethylene	16.3
MITC	100	0.5	6-mil polyethylene	5.3
Chloropicrin	17	not measured	Saran	3
			aluminum foil	0.2

[a] The flux density is: mg diffusing through 1 m^2 of film in 1 h while maintaining a 1 mg/L concentration gradient across the film.
[b] Hytibar® film is a high-barrier film manufactured in Belgium.

Traditional 1 mil (i.e., 0.025 mm) high-density polyethylene is relatively permeable to MeBr *(30,37)* and the permeability is affected by the ambient temperature. Using this material in warm temperatures can result in significant losses (i.e., 30 to 60%). However, under cool conditions and with a relatively deep injection depth, this plastic may provide adequate containment. Since experiments have shown that nearly all of the applied MeBr may leave the treated soil zone after a few days when injected at a shallow depth into bare soil *(12)*, under most circumstances it is better to use plastic than to leave the soil surface uncovered.

Table 3 provides a summary of the total MeBr emission in percent of applied MeBr for both tarped and untarped treatments following injection into soil columns (38). After correcting the flux for the presence of the lower boundary, the total emission loss of MeBr was 82% under bare surface conditions, and 43% under tarped surface conditions when injected at a 20-cm depth. For a 30-cm injection, 71% of the applied MeBr was emitted for the untarped column and 37% from the tarped column. When injected at 60-cm, the total emission loss was 38% under bare surface conditions, and 26% under tarped conditions.

Table 3. Effects of Injection Depth and Use of Plastic Films

Injection Depth	Total Emissions (Measured) (%)	Total Degradation (%)	Mass Balance (%)	Total Emissions Corrected Using Diffusion Model (%)
		Tarped Columns		
20 cm	59	36	94	43
30 cm	52	39	91	37
60 cm	45	46	91	26
		Non-Tarped Columns		
20 cm	91	12	102	82
30 cm	83	15	98	71
60 cm	60	36	96	38

When the soil surface was not covered with the polyethylene sheet, MeBr volatilization was extremely rapid, with as much as 80-90% of the total loss occurring during the first 24 h. In contrast, when a tarp was present on the soil surface, the maximum volatilization flux was significantly smaller, with only 30 - 45% of the overall loss occurring during the first 24 h. While measurable volatilization rates continued for a longer time (7-10 days) compared to the untarped columns (3-4 days), total emissions were significantly lower in tarped columns.

Similar results were observed in two parallel field experiments (12). In an untarped field, MeBr emission after 25-30 cm injection was measured to be 89% over the first 5 days after application; while in a tarped field located 6 km away, the emission rate was 32% over the first 9 days after application. Based on the results from this study and the few recently reported field studies, it is clear that MeBr emission rate in a tarped field is considerably lower than under untarped conditions when injected at shallow depth (20-30 cm). Films with lower permeability, such as Saranex®, should produce even greater reductions. Also, since MeBr is retained in the soil much longer under films with lower permeability, it should be possible to reduce the application rate without sacrificing the fumigation efficacy *(39)*. Reducing the application rate when high barrier films are used may provide a means for producing significant decreases in emissions from the combination of lower application rates and lower emissions.

Injection Depth. The depth of application is also an important factor affecting the amount of methyl bromide escaping into the atmosphere. In laboratory soil columns, when the application depth was increased from 20 to 60 cm, the MeBr emission rates decreased by 54% under untarped conditions, and 40% under tarped conditions (Table 3). The emission rate for the tarped, 60-cm application was the lowest estimated loss observed from any of the treatments (Table 3). This supports the results from a recent field experiment conducted at the Moreno Valley Field Station (Yates et al., 1994, unpublished data) where the MeBr emission rate for a bare soil, deep injection application was determined. MeBr was injected at approximately 68 cm at a rate of 322 kg/ha (291 lb/ac), a total mass of 1134 kg for the entire fumigated field. The average, maximum and minimum air temperatures during the first 7 days of this experiment were 15.1, 30.2, 4.5 °C, respectively. Shown in Figure 4 is the Br⁻ concentration reported as mass per sample length before and 3 and 9 months after application. From these data, it was estimated that 879 kg or (78%) of the applied mass was degraded to Br⁻; or approximately 21% of the applied mass was lost to the atmosphere. This is 66% less than the 62% total loss reported by Yates et al. *(11,13,16)* from an adjacent field and can be attributed to the deeper injection depth and a warmer average air temperature during the earlier experiment (24.2, 34.2 and 13.6 °C, for the 7 day average, maximum and minimum temperature).

These results are also in agreement with recent predictions *(35)* using a transport model. Under hypothetical conditions, it was estimated that increasing the

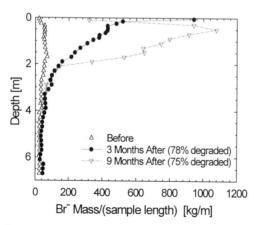

Figure 4. Bromide ion concentration as a function of depth in the field

injection depth from 25 to 45 cm would decrease the MeBr emission rates from 45 to 28% when the soil was tarped. From these findings, it can be concluded that placing

MeBr at a greater depth is another effective approach for minimizing its emission into the air during soil fumigation.

Soil Water Content. Increasing soil water content has been considered as a means for controlling MeBr movement $(33,35,37)$. The effect of water content on MeBr volatilization can be explained by the interactions of soil water content and the retardation factor, $R_d = (\theta + k_d \rho_b)/k_h + \varepsilon$, and tortuosity factor, (e.g., $\tau = (\phi - \theta)^{10/3}/\phi^2)$ in MeBr gas-phase transport, where ϕ, θ, ρ_b, k_d, k_h, respectively, are the porosity, water content, bulk density, liquid-solid and liquid-gas partition coefficients and $\varepsilon = \phi - \theta$. When the water content was increased from 0.058 to 0.180 cm³ cm⁻³, R_d increased from 1.21 to 1.58, τ decreased from 0.241 to 0.076 and the effective soil diffusion coefficient would be reduced by 76%.

In laboratory columns containing Greenfield sandy loam with 0.058 and 0.124 (cm³/cm³) volumetric water contents, the estimated loss after correcting for the presence of the column bottom was approximately 77% of the applied MeBr (See Table 4). When the water content was increased to 0.180 cm³ cm⁻³, only 62% was lost.

Table 4. Effects of Soil Type, Water Content and Bulk Density on MeBr Dissipation.

Treatment	Total Emissions (Measured) (%)	Total Degradation (%)	Mass Balance (%)	Total Emissions Corrected Using Diffusion Model (%)
Volumetric Water Content				
0.058	90	6	96	77
0.124	90	12	102	77
0.180	75	26	101	62
Bulk Density (g cm⁻³)				
1.40	90	12	102	77
1.70	64	29	93	53
Soil Type				
Greenfield SL	90	12	102	77
Carsetas LS	90	9	99	77
Linne CL	44	49	94	37

As the soil water content increased, the maximum MeBr flux density decreased and the time interval before reaching the maximum increased. Measurements of the MeBr gas concentration in the soil also indicated rapid movement through the soil column for the drier soils. MeBr in these soil columns was completely depleted 54 and 72 h after the application. For the wetter soil, measurable concentrations remained in the column until 144 h after the application.

In a recent field experiment (Yates et al., 1994 unpublished data), lower MeBr emissions were observed for bare soil, deep application than for a tarped, shallow application in the same field. Part of this difference may have been attributed to the water content of the soil profile. During the deep-injection study, the average soil water content around the injection point (68 cm below the surface) was 0.223 (cm^3/cm^3), whereas that observed during the shallow-injection study was 0.145 cm^3/cm^3. Yagi et al. *(15)* also attributed the decrease in MeBr emission from 87 in their first study to 34 % in their second study, in part, to soil moisture differences. Similar results were observed in the laboratory *(37)*.

Soil Bulk Density.   Soil bulk density can also have an effect on MeBr transport since the pore space decreases as bulk density increases. The bulk density, ρ_b, is related to the porosity from the relationship: *porosity = (1 - ρ_b/ρ_p)*, where ρ_p is the particle density.

In laboratory columns packed with Greenfield sandy loam, the corrected cumulative volatilization loss for a column with a bulk density of 1.70 g/cm^3 was 53%, significantly lower than the 77% loss from a column with a bulk density of 1.40 g/cm^3. The columns with higher bulk density behaved in a manner similar to the wetter soil column described above. Measurable volatilization continued for 120 h, the maximum flux density was reduced from 9.7 to 3.9 mg/h/column compared to the low bulk density column, and the time to reach the maximum flux increased from 2.5 to 6.5 h after application.

In the untarped, deep-injection field study (Yates et al., 1994, unpublished data), the field was disced and packed with a tractor shortly (approximately 5 min) after MeBr was injected into the field. The disking and surface packing closed the openings above the injection fractures and increased the bulk density near the surface. This, along with a higher water content, probably contributed to the reduced total emission compared to the shallow-tarped experiment. In practice, packing the soil surface and carefully closing the soil fractures created during application also should be considered for minimizing MeBr volatilization.

Degradation.   Along with volatilization of MeBr from the soil surface, hydrolysis and methylation are the principle degradative processes removing MeBr from agricultural soils *(40,41)*. Gan et al. *(41)* investigated the effect of soil properties on MeBr degradation and sorption in several soils and estimated the degradation half-life for MeBr in Greenfield sandy loam to be approximately 8 to 27 d, decreasing with increasing soil depth.

Degradation affects volatilization since it removes MeBr from the soil; making it unavailable for transport to the atmosphere. The effect of soil organic matter on MeBr volatilization has been investigated in our laboratory for three soils. The Greenfield sandy loam has relatively low organic matter and clay contents and is representative of many soil types in the state of California. Carsetas loamy sand has a very high sand content and very low organic matter and clay contents. Linne clay loam is relatively rich in organic matter and clay. Soil type had a pronounced effect on MeBr volatilization behavior as shown in Table 4. Volatilization of MeBr from untarped Carsetas and Greenfield soil columns following 30-cm injection was very rapid; both columns losing 77% of the applied MeBr. However, under the same conditions with the Linne clay loam, only 37% of the applied MeBr was lost. Analysis of Br⁻ concentration in soil at the end of the experiment revealed that 49% of the applied MeBr was degraded to Br⁻ in the Linne soil, while the degradation in Carsetas and Greenfield soils was approximately 10% (Table 4). The enhanced degradation of MeBr in Linne clay loam is likely due to its higher organic matter content *(41-43)*.

Using a gas-phase diffusion model, it was predicted *(35)* that when the soil organic carbon content was increased from 2 to 4%, the MeBr emission rate decreased from 45 to 37% following a tarped (2 days), 25-cm application under the assumed conditions. However, in his simulation, only the effect of soil organic matter on adsorption behavior was considered. From the column experiments, it is clear that enhanced degradation due to higher organic matter content may play an important role in reducing MeBr volatilization in organic-matter-rich soils.

Pesticide Efficacy. Efficacy and the rate of application are important factors in the fate and transport of MeBr used in pest control. If new management methods are developed which enhance MeBr efficiency, the quantity used in agricultural settings can be reduced resulting in less MeBr leakage into the atmosphere. To assure high efficiency of MeBr use, however, the uniformity and efficacy of the application must be determined.

Measuring MeBr concentrations in the soil gas phase at different depths provides some of the information needed to determine efficacy. Shown in Figure 5 are soil gas concentrations from the bare soil, deep injection experiment described earlier.

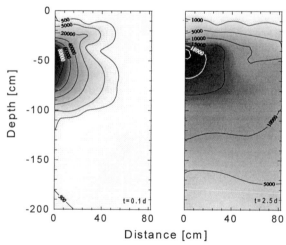

Figure 5. Soil gas concentrations at two times. Note: shading scales are different.

Initially, very high concentrations exist around the injection point. At later times, a more uniform concentration distribution occurs. This type of information is valuable to ensure that new management methods will be effective in controlling pests.

Deep placement of MeBr in coarse-textured soils is usually efficacious *(44,45)*. Application to the heavy-textured subsoil may be less effective, particularly if the soil is saturated at that depth. Therefore, the depth to which MeBr may be actually injected is dependent on soil conditions and the distribution pattern of target organisms, and should be decided by weighing between the efficacy and emissions under certain circumstances. In recent plot-scale experiments (Yates et al., 1995, unpublished data) MeBr gas (e.g. hot-gas method) was injected at 60 cm depth and covered with 1.4-mil polyethylene or 1-mil Hytibar® films to investigate how various management factors affect MeBr emissions. Located 4 cm deep in each plot were bags containing citrus nematodes, *Rhizoctonia solani* fungi and yellow nutsedge seeds. When the soil was covered with polyethylene, poor efficacy was observed in deep-injection plots. When covered with the high-barrier plastic listed in Table 2, good pest control was observed.

Conclusions

The great variation among results of recent experiments measuring the total emission of MeBr from fields imply that many factors, including those related to application methods as well as to soil and climatic conditions, integratively influence MeBr transport and transformation in the soil-water-air system and hence its ultimate loss from the soil surface. It was found that variables related to application methods, e.g., injection depth and use of surface tarp, and soil properties, e.g., water content, bulk density, soil organic matter have pronounced effects on MeBr volatilization following soil injection *(46,47)*.

The following conclusions can be drawn from this experimental information. Tarping consistently increased the residence time and amount of MeBr residing in the soil. The prolonged retention of MeBr in the soil resulted in more extensive degradation. Research indicates that the polyethylene film typically used for the surface cover is relatively permeable to MeBr and allows significant emissions compared to high-barrier plastic. This effect is more pronounced during periods of high temperature. Soil type, soil water content and bulk density are important factors affecting MeBr transport and transformation in soil, which ultimately affect volatilization. The total volatilization of MeBr from the organic-matter-rich Linne clay loam was only about half of that from a Carsetas loamy sand or a Greenfield sandy loam with relatively low organic matter contents. Organic matter additions which promote increased degradation offer another means for reducing volatilization. MeBr volatilization also decreased with increasing soil water content and bulk density. This dependence was mainly due to the reduced gas-phase diffusion as the result of reduced soil air porosity. Applying water at the soil surface can help to reduce volatilization losses.

To minimize volatilization, MeBr should be applied during periods of cool temperatures, relatively deep in organic-rich, moist soil under tarped conditions and the soil surface packed immediately after the application. Depending on site-specific conditions, a new high-barrier plastic should be used. Injecting MeBr during periods of warm temperature, at a shallow depth in dry, loose soil without the use of plastic barriers will likely result in maximum volatilization rates and, therefore, should be discouraged. Before adopting any new emission-reduction technology, the pest-control characteristics of the new methodology needs to be tested in typical regions, soils and environmental conditions. Failure to do this may produce unacceptable levels of pest control.

Acknowledgment. This study was supported by USDA Cooperative State Research Service Agreement No. 92-34050-8152.

Literature Cited

1. NAPIAP, 1993, The biologic and economic assessment of methyl bromide, United States Department of Agriculture, National Agricultural Pesticide Impact Assessment Program, Washington, D.C. April 1993, 99 pp.
2. Watson, R.T., D.L. Albritton, S.O. Anderson and S. Lee-Bapty, 1992, *Methyl Bromide: Its Atmospheric Science, Technology, and Economics*, United Nations Environment Programme (UNEP), United Nations Headquarters, Ozone Secretariat, P.O. Box 30552, Nairobi, Kenya, 1992
3. Wofsy, S.C., M.B. McElroy, and Y.L. Yung. 1975. The chemistry of atmospheric bromine. Geophys. Res. Lett. 2: 215-218.
4. Abritton, D.L., and R.T.Watson. 1992. Methyl Bromide Interim Scientific Assessment. United Nations Environment Programme, New York, Montreal Protocol Assess. Rep., 1992.
5. Singh, H.B. and M. Kanakidou. 1993. An investigation of the atmospheric sources and sinks of methyl bromide. Geophys. Res. Letters 20:133.
6. Khalil, M.A.K., R.A. Rasmussen, and R. Gunawardena. 1993. Atmospheric methyl bromide: trends and global mass balance. J. Geophys. Res. 98: 2887-2896.
7. Butler, J.H. 1995. Methyl bromide under scrutiny. Nature 376: 469-470.
8. Mano, S., and M.O.Andreae. 1994. Emission of methyl bromide from biomass burning. Science. 263:1255-1257.
9. Butler, J.H. 1994. The potential role of the ocean in regulating atmospheric CH_3Br. Geophys. Res. Lett. 21: 185-188.
10. Shorter, J.H., C.E. Kolb, P.M. Crill, R.A. Kerwin, R.W. Talbot, M.E. Hines, and R.C. Harriss. 1995. Rapid degradation of atmospheric methyl bromide in soils. Nature. 377:717-719.
11. Yates, S.R., J. Gan, F.F. Ernst, A. Mutziger, M.V. Yates, 1996a, Methyl Bromide Emissions from a Covered Field. I. Experimental Conditions and Degradation in Soil. J. Environ. Qual., 25:184-192.
12. Majewski, M. S., M.M McChesney, J.E. Woodrow, J.H. Pruger, and J.N. Seiber. 1995. Aerodynamic measurements of methyl bromide volatilization from tarped and nontarped fields, J. Environ. Qual. 24:742-752.

13. Yates, S.R., F.F. Ernst, J. Gan, F. Gao and M.V. Yates, 1996b, Methyl Bromide Emissions from a Covered Field. II. Volatilization. J. Environ. Qual. 25:192-202.

14. Yagi, K., J. Williams, N.Y. Wang and R.J. Cicerone. 1993. Agricultural soil fumigation as a source of atmospheric methyl bromide. Proc. Natl. Acad. 90:8420.

15. Yagi, K., J. Williams, N.Y. Wang and R.J. Cicerone. 1995. Atmospheric methyl bromide (CH_3Br) from agricultural soil fumigations. Science. 267:1979-1981.

16. Yates, S.R., J. Gan, F.F. Ernst, D. Wang, 1996c, Methyl Bromide Emissions from a Covered Field. III. Correcting chamber flux for temperature. J. Environ. Qual. 25(4).

17. Hollingsworth, E.B.. 1980. Volatility of trifluralin from field soil. Weed Sci. 28:224-228.

18. Clendening, L.D., W.A. Jury and F.F. Ernst. 1990. A field mass balance study of pesticide volatilization, leaching and persistence, Long Range Transport of Pesticides, D.A. Kurtz Ed., Lewis Publishers, Inc., Chelsea, Michigan, pp. 47-60.

19. Livingston, G.P, and G.L. Hutchinson, 1994, Enclosure-based measurement of trace gas exchange: Applications and sources of error, p. 14-51. *In* P.A. Matson and R.C. Harriss (ed.) Biogenic trace gases: Measuring emissions from soil and water, Blackwell Science LTD, Oxford, UK.

20. UNEP (United Nations Environmental Programme), 1992, MeBr science and technology and economic synthesis report, Robert T. Watson (Chair), 54 pp.

21. de Heer, H. Ph. Hamaker, L.G.M. Th. Tuinstra and A.M.M. van den Berg. 1984. Use of gas-tight plastic films during fumigation of glass-house soils with methyl bromide. Second International Symposium on Soil Disinfestation, Leuven, Belgium, September 26-30, 1983. Acta Horticulturae, 152:109-126.

22. Jury, W.A., 1985, Spatial variability of soil physical parameters in solute migration: A critical literature review, Research Report EA-4228, Electric Power Research Institute, Palo Alto, Ca, Sept. 1985, 78 pgs.

23. Parmele, L.H., E.R. Lemon and A.W. Taylor, 1972, Micrometeorological measurement of pesticide vapor flux from bare soil and corn under field conditions, Water, Air, Soil Pollution 1:433-451.

24. Yates, S.R., F.F. Ernst and W.F. Spencer, 1995, Design of a sampling mast for measuring volatile organic compounds in the near surface atmosphere, J. Environ. Qual. 24:1027-1033.

25. Gan, J., M.A. Anderson, M.V. Yates, W.F. Spencer, and S.R. Yates, 1995a, Sampling and stability of MeBr on activated charcoal tubes, J. Agric. Food Chem. 43:1361-1367.

26. Gan, J., S.R. Yates, W.F. Spencer and M.V. Yates, 1995b, Optimization of analysis of MeBr on charcoal sampling tubes, J. Agric. Food Chem. 43:960-966.

27. Wilson, J. D., G.W. Thurtell, G.E. Kidd, and E.G. Beauchamp, 1982. Estimation of the rate of gaseous mass transfer from a surface source plot to the atmosphere. Atmospheric Environment 16(8):1861-1867.

28. Wilson, J. D., V.R. Catchpoole, O.T. Denmead and G.W. Thurtell, 1983. Verification of a simple micrometeorological method for estimating the rate of gaseous mass transfer from the ground to the atmosphere. Agricultural Meteorology 29:183-189.

29. Denmead, O.T., J.R. Simpson and J.R. Freney, 1977, A direct field measurement of ammonia emission after injection of anhydrous ammonia, Soil Sci. Soc. of Am. J., 41: 1001-1004.
30. Kolbezen, M.J. and F.J. Abu El-Haj, 1977. Permeability of plastic films to fumigants, Proc. Int'l Agricultural Plastics Congress, San Diego, California, USA, April 11-16, 1977, pp. 1-6.
31. Schmidt, C.E. and W.D. Balfour, 1983, Direct gas emission measurement techniques and the utilization of emissions data from hazardous waste sites. Proceedings of the 1983 ASCE National Specialty Conf. on Environ. Eng., Boulder, CO, July 6-8, p 690-699.
32. Jury, W.A., W.F. Spencer and W.J. Farmer, 1983, Behavior assessment model for trace organics in soil: I. Model description, J. Environ. Qual., 12:558-564.
33. Goring, C.A.I., 1962, Theory and principles of soil fumigation, Adv. Pest Control Res. 5:47-84.
34. Kolbezen, M.J., D.E. Munnecke, W.D. Wilbur, L.H. Stolzy, F.J. Abu-El-Haj and T.E. Szuszkiewicz, 1974, Factors that affect deep penetration of field soils by MeBr, Hilgardia 42:465-492.
35. Reible, D.D., 1994, Loss of MeBr to the atmosphere during soil fumigation, J. Hazardous Materials, 37:431-444.
36. Massmann, J. and D.F. Farrier, 1992, Effects of atmospheric pressures on gas transport in the vadose zone, Water Resources Research. 28:777-791.
37. Jin, Y. and W.A. Jury. 1995. Methyl bromide diffusion and emission through soil columns under various management techniques, J. Environ. Qual., 24:1002-1009.
38. Gan, J.Y., S.R. Yates, W.F. Spencer, M.V. Yates and W.A. Jury, 1996, Effect of application methods on methyl bromide emission from soil. J. Environ. Qual. (in press)
39. Hamaker, Ph., H. de Heer, and A.M.M. van der Burg. 1983. Use of gas-tight films during fumigation of glasshouse soils with methyl bromide. II. Effects on the bromide-ion mass balance for a polder district. Acta Horticulturae 152: 127-135.
40. Gentile, I.A., L. Ferraris, M. Sanguinetti, M. Tiprigan and G. Fisichella, 1992, MeBr in natural fresh waters: hydrolysis and volatilisation, Pestic. Sci. 34:297-301.
41. Gan, J., S.R. Yates, M. Anderson, W.F. Spencer, and F. Ernst, 1994, Effect of soil properties on degradation and sorption of MeBr in soil, Chemosphere, 29:2685-2700.
42. Brown, B.D. and D.E. Rolston, 1980, Transport and transformation of methyl bromide in soils, Soil Sci., 130:68.
43. Arvieu, J. C., 1983, Some physico-chemical aspects of methyl bromide behavior in soil, Acta Horticulturae, 152: 267-275.
44. Lembright, H.W. 1990. Soil fumigation: principles and application technology. Suppl. J. Nematol. 22: 632-644.
45. Abdalla, N., D.J. Raski, B. Lear and R.V. Schmitt, 1974, Distribution of methyl bromide in soils treated for nematode control in replant vineyards, Pesticide Sci., 5:259-269.
46. Gan, J.Y., S.R. Yates, D. Wang, W.F. Spencer, and W.A. Jury. 1996. Effect of soil factors on methyl bromide volatilization after soil application. Environ. Sci. Tech. (in press)

Chapter 12

Error Evaluation of Methyl Bromide Aerodynamic Flux Measurements

Michael S. Majewski

U.S. Geological Survey, 2800 Cottage Way, Sacramento, CA 95825–1846

Methyl bromide volatilization fluxes were calculated for a tarped and a nontarped field using 2 and 4 hour sampling periods. These field measurements were averaged in 8, 12, and 24 hour increments to simulate longer sampling periods. The daily flux profiles were progressively smoothed and the cumulative volatility losses increased by 20 to 30% with each longer sampling period. Error associated with the original flux measurements was determined from linear regressions of measured wind speed and air concentration as a function of height, and averaged approximately 50%. The high errors resulted from long application times, which resulted in a nonuniform source strength; and variable tarp permeability, which is influenced by temperature, moisture, and thickness. The increase in cumulative volatilization losses that resulted from longer sampling periods were within the experimental error of the flux determination method.

Since the 1940's, agricultural pesticide use has been recognized as a potential source of atmospheric pollutants in terms of off-target drift of the applied materials. Drift occurs not only during the application process, but also from volatilization of the applied material. Volatilization continues as long as pesticides remain in the soil and a concentration-difference exists between the soil surface and the lower atmosphere. Volatilization also depends on the characteristics of the compound and its method of application. This post-application volatilization can be a significant source of pesticide input into the lower atmosphere that is now recognized as a major pathway by which pesticides can be transported and deposited in areas sometimes far removed from their sources (1).

An increasing number of environmental scientists, pesticide manufacturing companies, and state and local regulatory agencies are becoming aware that the atmosphere is an important component in the environmental fate of many pesticides. There is also increased interest in volatilization inputs into the lower atmosphere of these pesticides, especially methyl bromide.   Methyl bromide is used in the fumigation of harvested grains, fruits, and vegetables in stores, mills, and cargo ships, as well as in structural fumigation of buildings.   Methyl bromide is also used in agriculture.   It is injected into the soil as a pre-plant soil fumigant to control nematodes, fungi, weeds, and insects.   Typical application practices include bedding over the injection line or immediately laying a plastic tarp over the injection line. In California, there are also health concerns for communities surrounding fields where methyl bromide is used; to many farmers, however, methyl bromide is essential to their operation.

There is currently much concern and discussion focused on the continued use of methyl bromide because of its ozone depleting potential (*2, 3*).   Knowing how much of the applied material volatilizes into the lower atmosphere over a given time period is critical for estimating potential impact on ozone losses as well as downwind concentrations for estimating potential exposure and health effects.   Field volatilization flux data are also valuable in validating and fine tuning predictive environmental fate computer models.

The objective of this chapter is to discuss the statistical aspects of field volatility flux experiment measurements.   Specifically, how the length of sampling time affects the calculated volatility losses; the accuracy of the calculated flux measurements; and, does covering the field with a plastic film after a methyl bromide application reduce emissions into the atmosphere?

Measurement Methods

A variety of methods can be used to measure post-application volatilization of methyl bromide and other pesticides. These include vertical profile techniques, such as the Aerodynamic Gradient method (*4–6*) and an Integrated Horizontal Flux (*6, 7*) method; bulk transfer techniques, such as the Energy Balance Bowen ratio (*6, 8*) and Eddy Correlation (*6, 9*) methods; and trajectory simulations, such as the Theoretical Profile Shape method (*10, 11*).   Each of these methods and techniques have advantages and disadvantages that have been discussed in detail elsewhere (*6*).

Aerodynamic Gradient Method. One of the most frequently used field technique for measuring pesticide volatilization from treated surfaces is the aerodynamic gradient method. This method requires accurate vertical gradient measurements of wind speed, air temperature, and pesticide air concentration over a surface. The volatilization fluxes are calculated using the Thornthwaite-Holzman (*12*) equation modified to correct for atmospheric stability conditions (equation 1).

F is the vertical pesticide flux (mg m^{-2} h^{-1}), k is von Kármán's constant (dimensionless, $\approx$ 0.4), $\Delta\bar{c}$ (μg m^{-3}) and $\Delta\bar{u}$ (m s^{-1}) are the average pesticide air

$$F = \frac{k^2 \, \overline{\Delta c} \, \overline{\Delta u}}{\phi_m \, \phi_p \left[\ln \left(\frac{z_2}{z_1} \right) \right]^2} \tag{1}$$

concentration and wind speed differences between heights z_2 and z_1, respectively. ϕ_m and ϕ_p are diabatic functions that correct for atmospheric stability effects for pesticide and momentum, respectively. Equation 1 can be derived from the standard similarity form for the wind and air concentration profiles (*13*).

The universal phi functions (ϕ_m and ϕ_p) used in these experiments were developed by Pruitt *et al.* (*14*) and are dependent on the Richardson number (Ri), a stability parameter described by equation 2.

$$Ri = \frac{g(dT/dz)}{T(du/dz)^2} \equiv \frac{g \, \Delta T \, \Delta z}{T \, \overline{\Delta u}^2} \tag{2}$$

g (m s^{-2}) is gravitational acceleration, T (degree Kelvin) is the ambient air temperature, and dT/dz and du/dz are the air temperature and wind speed gradients, respectively. For unstable conditions (Ri < 0), which generally occur during daytime periods,

$$\phi_m = (1 - 16Ri)^{-0.33} \tag{3}$$

and

$$\phi_p = 0.885(1 - 22Ri)^{-0.40}. \tag{4}$$

For stable conditions (Ri > 0), which generally occur during calm, nighttime periods,

$$\phi_m = (1 + 16Ri)0.33 \tag{5}$$

and

$$\phi_p = 0.885(1 + 34Ri)^{0.40}. \tag{6}$$

The ϕ relationships correct only for thermal stability effects, not for the small effects on buoyancy due to water vapor, or for the potential error due to the inequality of the transfer coefficients.

The aerodynamic gradient, as well as many other field methods, depends on several requirements. The first is a large, spatially homogeneous surface source, typically 1-hectare or more. A sufficiently long up-wind fetch with similar topography surrounding the application area is needed to ensure that the boundary layer in which the fluxes are being determined has the same characteristic as the adjacent underlying surface. The assumption that steady-state conditions exist during the measurement periods is also made. These requirements and assumptions, together with the fact that wind speed and air concentration gradients are theoretically linear with the logarithm of the height, allow for the calculation of volatilization fluxes.

A typical aerodynamic gradient field experiment begins with the application of the compound or compounds of interest at a known rate (mass per area) to a field. The flux measurement equipment is usually set up in the middle of the field and sampling begins soon after the application is completed. The duration of sampling periods are variable and depend on the objectives of the experiment, trapping efficiency of the air-sampling matrix, and sensitivity of the analytical method.

Volatilization of the applied chemical can begin immediately, depending on the formulation and application method. In most cases with surface applied chemicals, the initial volatilization flux is very high and diminishes throughout the day because volatilization is a first-order, and therefore, exponential process. If the chemical is incorporated into the soil, or applied as a time-release formulation, the peak volatilization fluxes may follow a lag period of several hours to days. In general, volatilization flux follows diurnal cycles and are dependent on a variety of factors including solar input, soil moisture, and atmospheric stability (*15, 16*).

Effect of Sampling Time Length on Flux Results

Field volatilization flux experiments using short sampling periods of one to two hours or less generate hundreds of samples and are very labor intensive. Nevertheless, they produce a very detailed picture of the volatilization process. Not every field investigation is interested in such fine detail, however. Often, the required result is in terms of how much of the applied material was lost through volatilization over a given time period. One approach for reducing the field experiment work load is to use longer sampling periods – provided the sampling efficiency remains within acceptable limits – to calculate the resulting fluxes. As the sampling period increases, the detailed information about the diurnal volatilization cycles decreases. This, however, brings up important questions. Does the changing atmospheric stability over the longer sampling period affect the flux values? Are the cumulative volatilization losses for a field study that uses short sampling periods equivalent to the cumulative volatilization losses for a field study that uses longer sampling periods?

Field Experiment. The questions posed above can be answered by conducting a field experiment with several air sampling masts that sample for different lengths of time. This, however, is not practical because of the manpower, time, and expense involved. One solution is to take data from an experiment that utilized short sampling periods throughout, combine it at increasing time intervals, then compare the results. A 1992 field study investigating methyl bromide volatility provided results amenable to this type of manipulation. This experiment is described in detail elsewhere (*17*), and a brief description follows.

Methyl bromide was applied to two fallow fields approximately 6 km apart near Salinas, California, in October, 1992. It was injected into the soil 25 to 30 cm deep. One field was left open to the atmosphere and flux measurements were made for six consecutive days. The other field was covered with a high-barrier plastic tarp (0.00254 cm thick) during application and sampled for 10 consecutive days. The air was sampled at six levels with a maximum height of 2 m above the center of each field. Wind speed, air, and soil temperature gradients were also measured. The sampling periods for the first 24 hours were 2 hours followed by periods of 4 hours for the remainder of the experiment. Both fields exhibited high initial fluxes that declined throughout the first day. Peak daily fluxes for the remainder of the experiment generally occurred near mid-day and the lowest fluxes occurred late at night and early in the morning. Sampling over the tarped field was interrupted for 24 h beginning on day 5 because of rain. Sampling over the nontarped field was unaffected by the rain.

Volatilization Fluxes. The volatilization flux behavior over the fumigated field left open to the atmosphere is shown in Figure 1a. The square symbols represent the measured volatilization fluxes (mg m^{-2} h^{-1}) and the open circles represent the cumulative amount lost to the atmosphere by volatilization (kg ha^{-1}). Figure 2a shows the same information for the tarped field. The subsequent graphs (1b - 1d, and 2b - 2d) show the results when the data were combined into 8, 12, and 24 h sampling periods. Averaging the data into 4 h periods only affected the results for the first day's sampling of each field.

Flux values for the extended sampling periods were calculated using time weighted averages of the original air concentration and meteorological data for each extended time. For example, each 2 and 4 h value for wind speed, air temperature, differential air temperature, and air concentration that comprised an 8 h sampling period was multiplied by the original sampling time, summed together, then divided by the sum of the individual sampling time periods. The Ri, ϕ_m, and ϕ_p terms and the corresponding volatilization fluxes were then recalculated. Cumulative volatility loss was calculated by multiplying the calculated extended time period flux (mg m^{-2} h^{-1}) by the extended time period (h) and converting to the appropriate units (kg ha^{-1}). Both the combined period flux and the corresponding cumulative loss are plotted with respect to the midpoint of the averaged sampling time in Figures 1 and 2.

Discussion. Figures 1 and 2 illustrate with each progressively longer sampling time period, fine detail in the diurnal volatilization pattern is reduced and the volatilization flux profile for the length of the experiment is smoothed. Variations were due to gaps in the actual sampling which was not continuous. After the third day, sampling was interrupted from 10 a.m. to 12 p.m. at both fields. Sampling was also suspended between 1300 hours on 30 October to 1300 hours on 31 October over the tarped field due to heavy rain.

As the measured data were combined and fluxes were calculated for each extended time period, results were generally higher than the average of the original flux values. This was due to the atmospheric stability parameters, the Ri number and the $(\phi_m\phi_p)^{-1}$ term, in equation 1. The Ri values for both fields generally became more neutral; that is, as the time periods increased, the averaged Ri values approached zero. This resulted in an increase in the $(\phi_m\phi_p)^{-1}$ term for both fields, which increased the combined per period flux. This increase in the volatilization fluxes also increased the cumulative field losses by 28% (from 177 to 226 kg ha^{-1}) for the nontarped field,

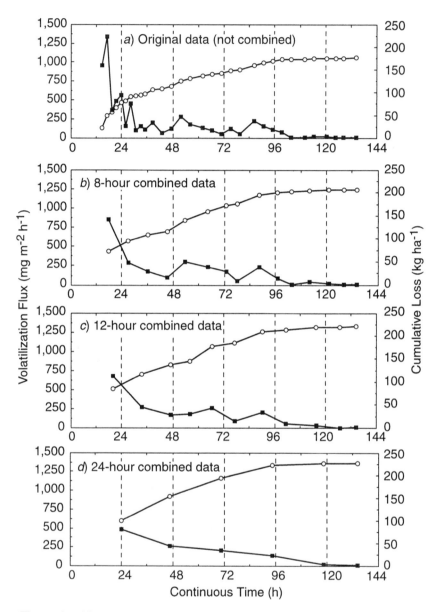

Figure 1.   Nontarped field volatilization fluxes. Original methyl bromide volatilization flux data (a) from the nontarped field. The symbol ■ represents the volatilization flux (mg m^{-2} h^{-1}) and O represents the cumulative loss (kg ha^{-1}). Figures b-d represent the 8, 12, and 24 hour combined average flux and cumulative loss calculated from the original data, respectively.

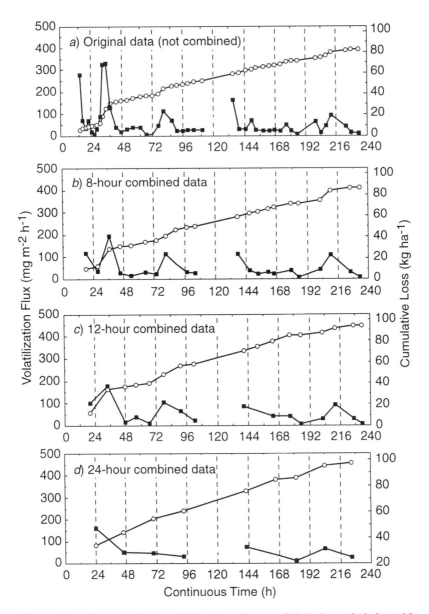

Figure 2.   Tarped field volatilization fluxes. Original methyl bromide volatilization flux data (a) from the tarped field. The symbol ■ represents the volatilization flux (mg m^{-2} h^{-1}) and O represents the cumulative loss (kg ha^{-1}). Figures b-d represent the 8, 12, and 24 hour combined average flux and cumulative loss calculated from the original data, respectively.

and by 19% (from 82 to 98 kg ha^{-1}) for the tarped field as the sampling periods were increased from 2 and 4 h to 24 h, respectively.

One question arose from the above results: are the increased cumulative volatilization losses resulting from the longer sampling periods significant? This can be addressed after the error associated with each flux value is determined.

Flux Error Determinations

Regression Analysis. To estimate volatilization fluxes using equation 1, accurate values of wind speed and air concentration are required. In this type of field experiment, however, one measurement is usually taken per height per sampling period and a variety of misfortunes can affect any sample at any time during the sampling and analytical process. One method that ensures the values used in equation 1 are as accurate as possible is regression analysis of the vertical gradient measurements (air concentration and wind speed). Since the air concentration and wind speed gradients are generally linear with respect to the logarithm of height above the surface, a linear regression of the best fit line through the data points provides a reliable way to estimate the $\Delta \bar{c}$ and $\Delta \bar{u}$ values needed in the volatilization flux calculations. In reality, these gradients are not always linear. Deviations from a linear profile typically only occur at the highest or lowest sampling point. This nonlinearity is due to an insufficient development of the overlying boundary layer, insufficient up-wind fetch, or interactions with the surface. Regression analysis of these data can still be done, provided the values at z_1 and z_2 fall within a linear portion of the gradient profiles. Since the values used in determining $\Delta \bar{c}$ and $\Delta \bar{u}$ are estimated from regression of the best fit line through the gradient data, the standard deviation associated with each estimated value can be calculated and an associated error assigned to the resulting flux values.

Determining errors associated with volatilization flux calculations begins by plotting the air concentration and wind speed measurements versus the logarithm of the sampling height. It is then visually inspected for linearity. Regression analysis is done using these data and the correlation coefficient (r) determined. If |r| is > 0.95, the error analysis can proceed. It should be noted that setting the critical value of |r| at 0.95 is arbitrary, and subject to the discretion of the analyst. If |r| is < 0.95, the plot must be reinspected to determine if the measured values bordering z_1 and z_2 are within a linear portion of the gradient profile. If so, the regression is redone using these data only. If the recalculated |r| is > 0.95, the error analysis can proceed. If |r| is < 0.95, the data can still be used if the trend in the data is nearly linear, but the resulting error will be higher. Next, the predicted mean values for wind speed and air concentrations at z_1 and z_2 are calculated from the equation of the fitted line. Occasionally, the vertical profiles of air concentration can be very erratic with no discernible trend. In these situations, volatilization fluxes usually cannot be determined with much confidence.

Error Estimation. The variance associated with the predicted mean value of the air concentration and wind speed estimates ($\bar{c}_1$, $\bar{c}_2$, $\bar{u}_1$, and $\bar{u}_2$) are determined next, percent uncertainty associated with $\Delta \bar{c}$ and $\Delta \bar{u}$, Ri, ϕ_m, and ϕ_p is then determined using the various equations for propagating uncertainty in a sum or difference, product or quotient, and power functions. These equations can be found in most introductory statistics books. This process can be simplified by assuming that T, ΔT, and Δz are measured accurately and their contributions to the overall error is negligible. Thus, the uncertainties in the Richardson number (Ri) and the stability corrections (ϕ_m and ϕ_p) are only dependent on the wind speed measurements. The percent uncertainty associated with the calculated fluxes can then be calculated along with the maximum and minimum range.

Confidence Limits. The variance of the predicted mean value of the wind speed and the air concentration values at z_1 and z_2 (est.$V(\hat{Y}_0)$), is determined using equation 7 (*18*)

$$est.V(\hat{Y}_0) = s^2 \left[\frac{1}{n} + \frac{(X_o - \overline{X})^2}{\Sigma (X_i - \overline{X})^2} \right] \qquad (7)$$

where n is the number of sampling heights, X_o in this experiment was log z_1 at 40 cm (1.602) or log z_2 at 140 cm (2.146), $\overline{X}$ is the average of the sampling heights, and X_i is the log of the actual sampling height. The mean about regression, s^2, is defined by

$$s^2 = \frac{SS}{(n - 2)} \qquad (8)$$

where n is the number of sampling heights and (n-2) is the degrees of freedom (d.f.). SS is the sum of the squares defined by

$$SS = \sum_{i = 1}^{n} (Y_i - \hat{Y}_i)^2 \qquad (9)$$

where Y_i is the measured wind speed or air concentration at z_i and $\hat{Y}_i$ is the corresponding estimated value at z_i. The standard deviation of the estimated value (est. s.d.($\hat{Y}_0$)) is determined using equation 10

$$est. \, s.d.(\hat{Y}_o) = \sqrt{est. \, V(\hat{Y}_o)} \qquad (10)$$

which is the square root of equation 7.

The confidence limits for the true mean value of Y ($\hat{Y}_o$) for a given X_o (in this case, the estimated air concentration and wind speed values at z_1 and z_2) are calculated as

$$\hat{Y}_o \pm t \; [est. \; s.d.(\hat{Y}_o)] \tag{11}$$

where t is the critical value at a chosen significance level, P (in this case P = 0.05) for (n-2) degrees of freedom. The t values are taken from the appropriate table of t-distributions found in most statistics books. With these data, the percent uncertainty in the $\Delta\bar{c}$ and $\Delta\bar{u}$ values can be determined.

Flux Value Uncertainty. Calculating the percent uncertainty associated with the per period volatilization flux is a step-wise process beginning with determining the percent uncertainty associated with the Ri, ϕ_m, and ϕ_p terms. The variables T, ΔT, and Δz in the Ri expression (equation 6) are assumed to have a negligible contribution to the overall error. Therefore, all the variables in equation 6 except $\Delta\bar{u}^2$ are treated as constants. The Ri expression is treated as a power function for the error calculation. The ϕ expressions are also treated as power functions with respect to Ri.

Table I shows the values and associated errors for all the variables used in the nontarped field flux calculations for the first day of the 1992 methyl bromide experiment (*17*). The uncertainties associated with each flux value from both fields are shown in Tables II and III, and graphically in Figures 3 and 4. When the errors associated with all the terms in equation 1 are known, the percent uncertainty in the final flux value can be determined.

Discussion. Methyl bromide is a highly volatile compound. It is applied as a liquid that volatilizes almost immediately upon injection into the soil. The application process during the 1992 experiment was long – 5 h for the tarped field and 4 h for the nontarped field. Air concentration measurements began immediately at the end of each application, but because methyl bromide is so volatile, the volatilization rate in the area where the application began was different than the rate in the area where the application ended. This produced a volatilization gradient over the field, which likely resulted in a nonuniform source during the first few sampling periods. It is also very likely that nonsteady-state conditions existed during the same periods. These conditions, together with residual airborne methyl bromide remaining from the application process, produced some erratic air concentration gradient profiles during the initial sampling periods that resulted in very large flux errors (Tables II and III, and Figures 3 and 4).

The errors in the measured fluxes during the first 2 periods for the nontarped field were much larger than for the tarped field (113% and 70% versus 46% and 36%, respectively). This difference was, most likely, due to nonsteady-state conditions brought on by the extreme volatility of methyl bromide, and by the regulating affect of the tarp itself that reduced the large initial flux from the tarped field. It is very likely that the 2-h sampling periods during the first day of the experiment were too long.

The overall averaged percent uncertainty in the volatilization fluxes from the nontarped field experiment was 44%, just slightly less than the 50% average

Table I. Variables and associated errors used to determine methyl bromide volatilization flux for the first day from the nontarped field

Date	Time		Log z (m)		T at 75 cm	
	On	Off	z_1	z_2	(°C)	dT/dz
27 October	1350	1608	1.48	1.90	16.85	-0.3425
	1615	1815	1.48	1.90	15.04	-0.0360
	1830	2038	1.48	1.90	14.88	0.1050
	2045	2245	1.48	1.90	14.72	0.1000
	2255	0055	1.48	1.90	14.49	0.1100

Date	Time		Wind Speed at 40 cm		Wind Speed at 140 cm		Uncertainty in ΔU	
	On	Off	U_1 (cm s^{-1})	s.d.	U_2 (cm s^{-1})	s.d.	cm s^{-1}	%
27 October	1350	1608	797	20.5	1030	25.4	32.7	14
	1615	1815	532	13.7	673	17.0	21.8	15
	1830	2038	254	6.41	326	7.96	10.2	14
	2045	2245	324	5.87	407	7.28	9.35	11
	2255	0055	336	5.69	421	7.06	9.06	11

Date	Time		Concentration at 40 cm		Concentration at 140 cm		Uncertainty in ΔC	
	On	Off	C_1 (mg m^{-3})	s.d.	C_2 (mg m^{-3})	s.d.	mg m^{-3}	%
27 October	1350	1608	2280	619	1390	774	991	111
	1615	1815	3250	833	1190	1072	1360	66
	1830	2038	2570	243	1340	288	377	31
	2045	2245	2310	522	892	671	851	60
	2255	0055	2610	575	1010	770	961	60

Date	Time		Ri	% Uncertainty in Ri	Φm	% Uncertainty in Φm	Φc	% Uncertainty in Φc
	On	Off						
27 October	1350	1608	0.0021	28.08	0.989	9.4	0.869	11
	1615	1815	-0.0006	30.83	0.997	10.3	0.880	12
	1830	2038	0.0068	28.15	1.035	9.4	0.962	11
	2045	2245	0.0049	22.51	1.026	7.5	0.942	9.0
	2255	0055	0.0052	21.29	1.027	7.1	0.944	8.5

Date	Time		Flux (mg m^{-2} h^{-1})	% Uncertainty in Flux	Flux at P = 0.05	
	On	Off			Maximum	Minimum
27 October	1350	1608	959	113	2040	-124
	1615	1815	1340	70	2270	406
	1830	2038	363	37	497	230
	2045	2245	494	62	801	188
	2255	0055	569	62	922	217

% uncertainties are for mean values. s.d., standard deviation

Table II. Hourly volatilization flux values and associated maximum and minimum error values with corresponding cumulative loss for the nontarped field

Date	Midpoint Cumulative Sampling Time (h)	Flux $(mg\ m^{-2}\ h^{-1})$	% Uncer- tainty	Flux $(mg\ m^{-2}\ h^{-1})$ Maximum	Minimum	Cumulative Loss $(Kg\ ha^{-1})$
27 October	14.98	959	113	2040	-124	22.1
	17.25	1340	70	2270	406	48.8
	19.57	363	37	497	230	56.6
	21.75	494	62	801	188	66.5
	23.92	569	62	922	217	77.9
28 October	26.10	158	81	286	80.5	81.0
	28.50	460	52	700	220	90.2
	30.75	98.3	43	140	56.2	92.2
	33.00	152	43	217	87.2	95.2
	35.25	115	54	177	52.4	97.5
	38.50	200	40	281	119	106
	43.04	63.5	21	76.6	50.5	108
	47.33	124	43	178	71.2	113
29 October	51.69	285	19	338	232	125
	56.00	182	73	315	48.9	132
	62.25	134	32	177	90.8	137
	66.58	97.9	26	123	72.9	141
	71.02	49.8	21	60.0	39.6	143
30 October	75.33	120	37	164	75.2	148
	79.67	50.1	29	64.4	35.7	150
	86.25	228	27	290	167	159
	90.58	161	23	197	124	166
	95.04	108	27	137	78.7	170
31 October	99.33	72.3	14	82.5	62.2	173
	103.67	8.72	18	10.3	7.11	174
	110.27	7.94	15	9.11	6.76	174
	114.58	16.7	10	18.4	15.0	175
	120.30	16.8	28	21.4	12.2	176
1 November	124.88	11.2	181	31.3	-8.99	176
	128.25	3.22	39	4.49	1.96	176
	134.25	11.0	29	14.2	7.79	176

Table III. Hourly volatilization flux values and associtated maximum and minimum error values with corresponding cumulative loss for the tarped field

Date	Midpoint Cumulative Sampling Time (h)	Flux (mg m^{-2} h^{-1})	% Uncertainty	Flux (mg m^{-2} h^{-1}) Maximum	Minimum	Cumulative Loss (Kg ha^{-1})
26 October	15.00	279	46	408	151	5.59
	17.28	70.4	36	96.0	44.7	7.03
	19.52	35.1	29	45.2	25.0	7.74
	21.78	69.1	17	80.7	57.5	9.17
	24.11	19.2	171	52.0	-13.6	9.58
27 October	26.33	7.48	57	11.8	3.20	9.73
	28.68	29.0	53	44.4	13.6	10.4
	30.94	88.3	46	129	48.0	12.1
	33.33	325	41	457	193	18.6
	35.67	329	24	408	250	25.9
	39.02	130	35	175	85.2	31.0
	43.31	37.7	37	51.5	23.8	32.6
	47.60	17.8	79	31.9	3.72	33.3
28 October	51.92	28.7	**	----	----	34.6
	56.52	38.7	57	60.7	16.6	36.3
	63.08	39.7	23	48.9	30.4	37.9
	67.60	6.04	68	10.2	1.92	38.2
	72.00	5.27	58	8.33	2.20	38.4
29 October	76.37	44.8	57	70.3	19.4	40.2
	80.78	112	25	141	83.8	44.8
	87.00	71.2	48	105	36.9	47.6
	91.33	21.3	19	25.4	17.3	48.5
	95.71	22.2	81	40.2	4.19	49.4
30 October	100.00	25.2	47	37.0	13.4	50.4
	104.33	26.6	29	34.4	18.8	51.5
	111.15	25.9	**	----	----	52.5
31 October	135.25	166	25	207	125	59.1
	139.57	28.7	**	----	----	60.3
	145.17	30.5	118	66.5	-5.52	62.3

**, Air concentration data for these periods were lost. Flux values were calculated as the time weighted average of the preceding and following period fluxes.

continued on next page

Table III. Hourly volatilization flux values and associtated maximum and minimum error values with corresponding cumulative loss for the tarped field—*Continued*

Date	Midpoint Cumulative Sampling Time (h)	Flux $(mg\ m^{-2}\ h^{-1})$	% Uncertainty	Flux $(mg\ m^{-2}\ h^{-1})$ Maximum	Minimum	Cumulative Loss $(Kg\ ha^{-1})$
1 November	149.67	69.2	84	128	10.9	63.8
	152.92	25.6	111	54.0	-2.77	64.8
	159.04	23.2	17	27.2	19.3	65.7
	163.42	21.7	59	34.3	8.97	66.6
	167.71	25.2	53	38.5	11.8	67.6
2 November	172.08	20.9	54	32.1	9.68	68.6
	176.67	50.8	124	114	-12.4	70.6
	181.00	21.8	56	34.2	9.53	71.5
	185.12	6.15	41	8.70	3.61	71.7
3 November	200.13	64.3	34	85.8	42.8	74.2
	204.00	13.4	57	20.9	5.81	74.7
	207.79	44.0	55	68.1	19.9	76.4
	211.87	94.0	51	141.6	46.4	80.3
4 November	224.00	40.7	26	51.2	30.1	81.7
	227.83	15.4	73	26.6	4.12	82.3
	233.42	9.12	24	11.3	6.95	82.4

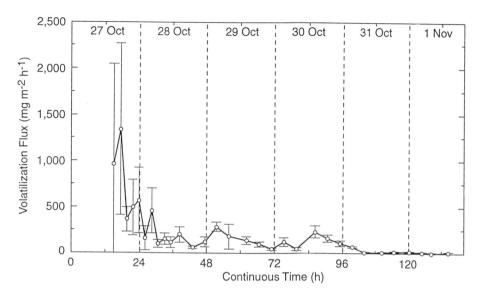

Figure 3. Methyl bromide volatilization flux per period with associated error for the nontarped field.

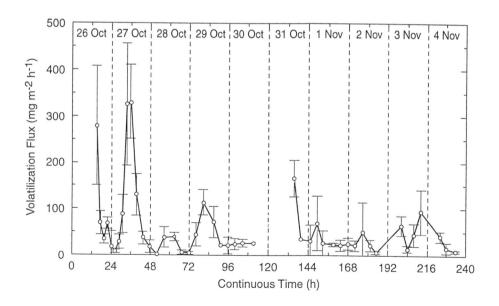

Figure 4. Methyl bromide volatilization flux per period with associated error for the tarped field.

uncertainty for the tarped field experiment. The averaged daily percent uncertainty in the volatility flux measurements for the nontarped field, with the exception of the last day, however, decreased with time from 70% to 17%. This was not the case for the tarped field where the averaged daily percent uncertainties in the flux measurements were variable, but did not exhibit any downward trend. This indicates that while the tarp retarded methyl bromide emissions, it also affected the quality of these emissions over the treated area.

Tarp Effects. The sustained high uncertainties associated with the averaged daily volatilization fluxes off the tarped field were most likely due to the regulating affect of the tarp itself. The tarp contained the methyl bromide in the soil. It is permeable however, and the rate at which methyl bromide permeates it is a function of film thickness and density.   The thickness and density of the tarp is specified by the manufacturer but it does vary (19). In addition, each successive tarp swath was overlapped and glued, creating narrow double-layer strips which added to the inconsistent film thickness and film permeability throughout the application area. This resulted in areas of variable methyl bromide emissions, which affected the uniformity of the source, and possibly, the steady-state conditions above the tarp, both of which can affect the uncertainty in the flux measurements. Water also can decrease the permeability of the tarp. A 1-mm thick layer of water can decrease the permeability by 15% (20). The rain event, which occurred on day five of the experiment and interrupted sampling at the tarped field for about 24 h, left areas of water pooled on the tarp. Temperature has a direct effect on the tarp permeability too (20). The highest fluxes usually occurred during the middle of the day when air temperatures were highest, and the lowest fluxes usually occurred at night when air temperatures were lowest (Figure 4).   The overall effect of the variable tarp permeability was to decrease the uniformity of the source strength and disturb the steady-state conditions over the tarp during the sampling periods. This resulted in greater variability in the measured air concentration gradients that increased the associated error in the estimated averaged air concentration values used in the flux calculations.

High errors associated with the air concentration measurements can also be attributed to the fact that only six sampling levels were used. The error estimation method requires (n-2) degrees of freedom (d.f.). Using only 6 sampling points resulted in a large critical value (t) multiplier in the determination of confidence limits (2.776 for [n - 2] = 4, and 3.182 for [n - 2] = 3). The critical value decreases as the number of sampling points increases, therefore, the confidence in predicting error limits can be increased by increasing the number of gradient measurement points. Reducing the significance level from P = 0.05 to P = 0.10 or lower will also reduce the critical value multiplier in the confidence limit determinations. This should be done only if the greater risk of being wrong is acceptable.

The average error associated with the volatilization flux values from both fields was about 50%, but are the results from the two fields significantly different? Does covering the field with plastic film reduce methyl bromide emissions into the atmosphere? These questions were addressed by comparing the flux results from both fields using a paired t-test.

Paired t-test. The flux results for the first 5 days of both field experiments were compared using a paired t-test to determine if the respective volatilization losses were significantly different. When the results were compared on a daily basis, only 2 of the 5 days showed a significant difference at the P = 0.05 level. These were the first and third days. On the first day, the extremely high volatilization fluxes from the nontarped field – nearly five times greater than those from the tarped field – are the most likely explanation for the difference. On the third day, the very low fluxes from the tarped field are the likely reason for the difference. Fluxes from both fields were nearly equivalent on the second day. On the fourth and fifth days, the fluxes over the nontarped field rapidly decreased to nearly the same levels as the tarped field. When the results for the entire five days were compared however, the results indicated that there was an overall significant difference in volatilization fluxes between the two fields at the P = 0.05 significance level, with the nontarped field having the higher flux. This was expected since the nontarped field had virtually nothing restricting the volatilization of the applied methyl bromide. The nontarped field lost greater than 50% of the nominal application during the first 24 hour compared to only 10% from the tarped field.

The cumulative volatilization losses were also compared using a paired t-test and the results showed a significant difference in daily losses between the two fields at the 95% confidence interval (177 kg ha^{-1} for the nontarped field versus 82 kg ha^{-1} for the tarped field). These results indicate that covering a field with a plastic tarp after a methyl bromide application significantly lowers volatilization losses compared to a nontarped application for the same time period.

With an average volatilization flux measurement error of 50% from both the tarped and nontarped fields, the 20 to 30% increase in cumulative losses resulting from fluxes calculated using 8, 12, and 24 h sampling periods falls within the experimental error of the aerodynamic method, and is not statistically significant in this experiment.

Field experiments measuring the volatilization flux of various pesticides have been conducted since the 1960's (*1, 4, 21*). A variety of methods have been used, but an underlying question that has troubled many researchers is the accuracy of the measured flux values. The results presented here indicate that with methyl bromide, the average uncertainty in the volatility flux measurements is about 50%. This represents a likely upper limit of measurement error. One reason is the vapor pressure of methyl bromide. It is many times greater than most pesticides in use today and it results in extremely high volatilization rates. These high volatilization rates can produce a nonuniform source and nonsteady-state conditions. The extremely high initial volalatilization rates, especially from the nontarped field results in unaccounted methyl bromide volatilization losses that occur during the application process and before measurements begin. Many pesticide field volatilization flux experiments are also conducted over fallow fields without the complication of a permeable plastic tarp covering the soil.

Conclusions

Methyl bromide volatilization fluxes were calculated using air concentration and wind speed values obtained from linear regression of measured gradients with height over a nontarped and a tarped field for 6 and 10 days, respectively. The field measurements were averaged in 8, 12, and 24 h time periods to simulate sampling periods longer than the original 2 and 4 h sampling periods. The results of this simulated increase in sampling time indicate fluxes calculated using time-weighted averages of measured air concentration and meteorological data were generally higher than a simple average of the original per period fluxes. The cumulative losses also increased by 20 to 30%, primarily due to the neutralizing affects of the averaged atmospheric stability terms in the flux calculations.

The error associated with each volatilization flux measurement was determined by calculating the percent uncertainty associated with the air concentration and wind speed data used in the flux calculation and propagating this error to determine the percent uncertainty associated with the flux value. The associated error for each flux measurement averaged about 50%. Reasons for the high errors include long application times that resulted in a nonuniform source strength, nonsteady-state conditions that resulted from rapidly volatilizing methyl bromide, and variability in the tarp permeability, which was affected by temperature, moisture, and variable thickness. Volatility fluxes and cumulative volatilization losses were compared and found to be significantly different at the $P = 0.05$ significance level. The errors associated with longer sampling periods were within the experimental error of the flux determination method.

Although using time weighted averages of measured variables from short sampling periods to simulate results for longer sampling periods may not precisely duplicate the results from an actual field experiment, it does provide examples of the most likely trends to expect. Additional field experiments, however, need to be conducted to verify that these trends do indeed approximate what is occurring over the longer sampling intervals.

Literature Cited

(1) Majewski, M.S.; Capel, P.D. *Pesticides in the Atmosphere: Distribution, Trends, and Governing Factors*; Gilliom, R.J., Ed.; Pesticides in the Hydrologic System; Ann Arbor Press, Inc.: Chelsea, MI, 1995; Vol. 1; 250 p.

(2) Methyl Bromide Science and Technology and Economics Synthesis Report; Watson, R.T., Ed; Montreal Protocol Assessment Supplement; United Nations Environment Programme on behalf of the Contracting Parties to the Montreal Protocol: 1992, 33 p.

(3) *Annual International Research Conference on Methyl Bromide Alternatives and Emissions Reductions Proceedings*; Sponsored by Methyl Bromide Alternatives Outreach in cooperation with U.S. Environmental Protection Agency and U.S. Dept. of Agriculture, 1995, 246 p.

(4)   Parmele, L.H.; Lemon, E.R.; Taylor, A.W.   *Water, Air, Soil Poll.* 1972, *1*, 433-451.
(5)   Shewchuk, S.R.; Grover, R.   *Description of a portable micrometeorological system to study pesticide volatilization and vapour transport*; SRC Technical Report No. 123; Saskatchewan Research Council: Regina, Saskatchewan Canada, March, 1981, 37 p.
(6)   Majewski, M.S.; Glotfelty, D.E.; Paw U, K.T.; Seiber, J.N.   *Environ. Sci. Technol.* **1990**, *24*, 1490-1497.
(7)   Denmead, O.T.; Simpson, J.R.; Freney, J.R.   *Soil Sci. Soc. Amer. J.* **1977**, *41*, 1001-1004.
(8)   Bowen, I.S.   *Phys. Rev.* **1927**, *27*, 779-787.
(9)   Majewski, M.S.; Desjardins, R.L.; Rochette, P.; Pattey, E.; Seiber, J.N.; Glotfelty, D.E.   *Environ. Sci. Technol.* **1993**, *27*, 121-128.
(10) Wilson, J.D.; Catchpoole, V.R.; Denmead, O.T.; Thurtell, G.W.   *Agric. Meteorol.* **1983**, *29*, 183-189.
(11) Majewski, M.S.; Glotfelty, D.E.; Seiber, J.N.   *Atmos. Environ.* 1989, *23*, 929-938.
(12) Thornthwaite, C.W.; Holzman, B.   *Mon. Weather Rev.* **1939**, *67*, 4-11.
(13) Monin, A.S.; Obukhov, A.M.   *Tr. Geofiz. Inst. Akad. Nauk. S.S.S.R.* **1954**, *24*, 163-187.
(14) Pruitt, W.O.; Morgan, D.L.; Lourence, F.J.   *Q. J. Roy. Meteorol. Soc.* **1973**, *99*, 370-386.
(15) Spencer, W.F.; Cliath, M.M.   In *Long Range Transport of Pesticides*; Kurtz, D.A., Ed.; Lewis Publishing Co.: Chelsea, MI; 1990, pp 1-16.
(16) Spencer, W.F.; Farmer, W.J.; Jury, W.A.   *Environ. Toxocol. Chem.* **1982**, *1*, 17-26.
(17) Majewski, M.S.; McChesney, M.M.; Woodrow, J.W.; Prueger, J.H.; Seiber, J.N.   *J. Environ. Qual.* **1995**, *24*, 742-752.
(18) Draper, N.; Smith, H.   *Applied regression analysis*; John Wiley & Sons, New York; 1981; 709 p.
(19) Yagi, K.; Williams, J.; Wang, N.Y.; Cicerone, R.J.   *Proc. Natl. Acad. Sci.* **1993**, *90*, 8420-8423.
(20) Kolbezen, M.J.; Abu-El-Haj, F.J.   *Permeability of plastic films to fumigants*; In Proc. 7th International Agricultural Plastics Congress, San Diego, CA, 11-16 April, 1977; International Agricultural Plastics Congress, Chula Vista, CA, 1977, pp 476-481.
(21) Caro, J.H.; Taylor, A.W.; Lemon, E.R.   *Measurement of pesticide concentrations in the air overlying a treated field*; In Proc. Int. Symp. Identification, Measurement Environ. Pollutants, Ottawa, ON Canada, **1971**, pp 72-77.

Chapter 13

Flux, Dispersion Characteristics, and Sinks for Airborne Methyl Bromide Downwind of a Treated Agricultural Field

James N. Seiber[1], James E. Woodrow[1], Puttanna S. Honaganahalli[1], James S. Lenoir[1], and Kathryn C. Dowling[2]

[1]Center for Environmental Sciences and Engineering and Department of Environmental and Resource Sciences, Mail Stop 199, University of Nevada, Reno, NV 89557–0187
[2]Department of Environmental and Occupational Health, School of Public Health, Loma Linda University, Loma Linda, CA 92354

A field study was conducted of methyl bromide volatilization flux, dispersion, and atmospheric fate in Monterey County, California, in 1994. Air concentrations of methyl bromide were measured above a fumigated field and downwind from the field with the objective of comparing vertical flux with horizontal flux based upon measured methyl bromide concentration and meteorological data. Another objective was to compare downwind air concentration data to concentrations predicted by a dispersion model (Industrial Source Complex-Short Term II [ISC-STII] model) in order to assess the value of this model for use in exposure assessment for agricultural workers and downwind residents. Flux, vertical profile, and single-height air concentrations were measured at four different locations extending from the center of the treated field to nearly 0.8 km downwind. These concentrations were used to determine air concentration profiles and downwind dispersion concentrations. The dispersion values were less than those generated from the ISC-STII Model by a factor of ~2. This study showed (1) Average horizontal flux near the downwind edge was ~99% of mid-field vertical flux; (2) Vertical air concentration and flux profiles farther downwind showed depletion of methyl bromide near the surface of an adjoining mature strawberry field. Adsorbed methyl bromide was found in the surface soil of the same field; and (3) Methyl bromide air concentrations measured downwind of the treated field source compared with those predicted by ISC-STII model, allowing the construction of downwind isopleths for each day of sampling.

Agriculture represents an important anthropogenic source of methyl bromide emissions to the atmosphere. Its use as a pre-plant soil fumigant for the control of nematodes, soil-borne pathogens, weeds, and other biological pests accounts for about 80% of all methyl

0097–6156/96/0652–0154$16.00/0

bromide use worldwide (about 63 million kg in 1990 *(1)*). It has been shown that about one-third, and perhaps more, of the one million-plus kg of methyl bromide applied as a soil fumigant in California volatilizes, even when it is injected into the soil and confined with a plastic tarp *(2, 3)*. Partly because of this, there has been a growing concern over the role methyl bromide may play in stratospheric ozone depletion. This concern is based on the fact that bromine, one source of which is the photochemical conversion of methyl bromide, has been shown to be about 40 times more effective than chlorine in ozone depletion potential *(4)*. As a result, methyl bromide has been categorized as a Class 1 ozone-depleting chemical by the U.S. EPA, and the United Nations Environmental Program has called for at least a 25% reduction in its worldwide use, if not an outright ban, by the year 2010 *(5)*. The U.S. EPA, under rules in the Clean Air Act Amendment (1990), has also mandated a ban on the use of methyl bromide as a soil fumigant.

What is often not considered in discussions concerning methyl bromide is the possible role 'sinks' might play in removing this material from the atmosphere and preventing its eventual movement into the stratosphere. Possible sinks include (1) chemical/photochemical breakdown; (2) wet deposition in rain and fogwater; and (3) dry deposition of vapors to soil and plant surfaces. Other investigators have provided evidence that possibility (1) is negligible *(6)*, at least in localized application-release situations. Possibility (2) may be important during periods of rain or fog, but these tend to be intermittent and infrequent in many areas where methyl bromide is used in agriculture. This field study was originally designed to help discern the importance of (3). If experimental concentrations of methyl bromide downwind of a treated field were found to be lower than predicted by a dispersion model (e.g., ISC-STII) which has no sinks in it, breakdown and/or dry deposition of methyl bromide might be indicated. A further indication of exchange with a surface might be from depletion of methyl bromide in air near the surface *(7)*, while the presence of methyl bromide in the downwind soil surface could provide supportive evidence for dry vapor deposition exchange. In order to evaluate loss of methyl bromide from the downwind air mass, sampling was conducted with flux masts placed at several locations, including near the middle of the field at the downwind edge, and at several additional downwind locations. Single height sampling masts were also positioned at strategic sites as far away as 0.8 km.

Although this study did not provide definitive information on the operation of depositional sinks, it did allow for a comparison between aerodynamic (vertical) flux and integrated horizontal flux measurements to better characterize the plume source and it provided an opportunity to evaluate the applicability of ISC-STII to predicting air concentrations at various downwind distances. Additionally, some evidence was obtained indicating that deposition of methyl bromide vapors to soil was occurring; this possibility can be further explored using some of the methods described here.

Experimental Procedure

Site Location and Treatment. On August 12, 1994, Tri-Cal, Inc. (Hollister, CA) fumigated a 15.6 ha Salinas Berry Farm strawberry field located at 397 Natividad Road in Monterey County. The treated field was roughly quadrilateral in shape with a 433-meter base on the eastern edge and a slightly curved western side that followed a dry creek bed (Figure 1). Tri-Cal injected 6,118 kg (for an application rate of 392 kg/ha) of Tricon 67/33 (67% methyl bromide/33% chloropicrin) to a depth of 25 cm and

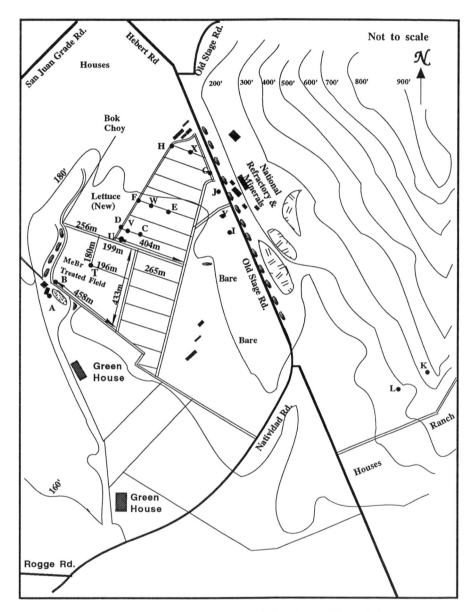

Figure 1.   Map of methyl bromide-treated field showing position relative to sampling locations and other fields.

concurrently tarped the field with 1 mil polyethylene strips which were sealed at the seams -- all in accord with local fumigation practices. The field tarp was cut five days after application, on August 17, 1994, and the tarp was removed on August 18, 1994.

Sampling Masts. Air sampling stations were established upwind of the treated field, in the center of the field, and at various distances downwind. The placement of the sampling stations was primarily dictated by such conditions as terrain, wind direction, and accessibility. Wind direction was mainly from the west and southwest in the afternoons. Morning winds were variable, generally from the west but occasionally from the east. Our experimental design incorporated a mast near the center of the field (T) for in-field flux; three masts (U, V, W) for downwind vertical profiles; duplicate long-range air samplers (A-L) and duplicate triple-height long-range air samplers (X, Y) to compensate for changes in wind direction across the treated field (Table I). The sampler locations are shown in Figure 1.

The approximate center of the field (location T) was identified by measuring 180 meters into the field from the center of the northeastern side and 196 meters into the field from the mid-point of the southeastern edge. Two identical masts with six sampling slots each were interchanged from one sampling interval to the next at location T to measure vertical flux via the aerodynamic gradient method *(3, 8)*. Mast T sampling heights were approximately 15, 30, 45, 80, 125, and 200 cm. This mast, located in the center of the treated field, was accompanied by five Met One (Grants Pass, OR) Model 5341 anemometers mounted on a mast at 9, 28, 77, 128, and 208 cm. Five Campbell Scientific (Logan, UT) TCBR-3 thermocouple probes located at 8, 30, 76, 126, and 208 cm were used to measure temperature gradients.

Mast U served as an integrated horizontal flux mast *(9, 10)*. This mast was placed 6 m off of the downwind edge of the treated field (187 meters northeast of the center of the field). Another mast (Mast V) was located further downwind (281 meters northeast of the center of the field). An anemometer mast adjacent to mast U was equipped with six Model 901-LED Sensitive Cup Anemometers (C.S. Thornthwaite Associates, Elmer, NJ) at 12, 30, 60, 100, 150, and 200 cm and a Met One Model 023/024 Wind Direction Sensor at 2.6 m. Mast V was accompanied by a similar anemometer mast with six Model 901-LED Sensitive Cup Anemometers at 14, 30, 60, 100, 150, and 200 cm and a Met One Wind Direction Sensor at 2.6 m. A further mast, Mast W, with samplers at five heights, was located 476 meters northeast of the center of the field. Nearby, a Young (Traverse City, MI) Model 05305 Wind Monitor-AQ continually measured wind direction and wind speed. Concentration and wind speed data from Mast U was used to calculate integrated horizontal flux (IHF) *(9, 10)*:

$$IHF = \frac{1}{X} \int_0^z \overline{c}_i \; \overline{u}_i \; dz$$

where c_i and u_i are average concentration and wind speed, respectively, at height z_i, z is plume height, and X is the distance to the upwind edge of the source.

Fourteen two-meter high duplicate air sampling masts (A-L) were placed at

Table I. Methyl bromide sampling mast types and locations

Mast Type	Sampling heights, cm	Location
A-L	200	see Figure 1
T	15, 30, 45, 80, 125, 200	center of treated field
U	12 or 15, 30, 60, 100, 150, 200	187 m downwind of field center (downwind edge)
V	12 or 15, 30, 60, 100, 150, 200	281 m downwind of field center
W	20, 60, 120, 200, 400	476 m downwind of field center
X, Y	200, 400, 600	see Figure 1

various locations both upwind and downwind of the treated field. Masts A-L, although not accompanied by stationary weather stations, were monitored with Turbometer portable anemometers (Davis Instruments, Hayward, CA) and Fisherbrand hand-held digital thermometers (Fisher Scientific, Pittsburgh, PA) as samples were installed and removed. In addition, surveyor ribbons tied to the top of each mast were used along with Silva (LaPorte, IN) compasses to measure wind direction at the start and finish of each sampling period.

Two final air sampling masts, Masts X and Y, with duplicate samplers at heights 2, 4, and 6 m, were installed at two different locations downwind of the treated field. For Mast Y, wind speed, temperature, and wind direction were monitored as described for Masts A-L. However, Mast X was located adjacent to a second Young Model 05305 Wind Monitor-AQ that continuously monitored wind direction and wind speed.

Meteorological Data. All Campbell Scientific, Thornthwaite, Met One, and Young meteorologic equipment was connected to Campbell Scientific dataloggers placed at locations T, U, V, W, and X. Weather data were collected at intervals of 5 minutes through all of the days of the study. Dataloggers were downloaded daily, and the collected data were imported into Microsoft Excel (Version 5.0) for processing. Measured temperatures ranged from 7 to 30°C over the course of the study. Wind speeds in the vicinity of the treated field reached 6 m/sec.

Sampling Schedule. Prior to field treatment, duplicate background samples were taken at the side of the field, near the future site of Mast U. These background samples were taken at a height of 1.5 meters on the evening of August 11, 1994. Sampling of the treated field continued from August 12-18, 1994. Several additional samples were taken on September 9 and 10, 1994. Table II summarizes the samples collected at various sampling mast locations over the course of the study.

Sampling Protocol. Sampling trains consisted of primary and secondary air sampling tubes connected in series to a vacuum source. Each glass sampling tube was commercially packed with approximately 1 gram of Lot 120 coconut charcoal by SKC-West (Fullerton, CA). The sealed tubes were opened by clipping the ends with needle-nose pliers and a sampling train was prepared by connecting two open tubes in series using a small piece of Tygon tubing, making sure that the two adjacent ends were as close together as possible. Stations U, V, K, and L were connected to Staplex (Brooklyn, NY) high volume air samplers through a manifold adaptor. Optimum flow rates and sampling intervals were 100 mL/min for four hours (approximately 24 liters of air), although actual recorded flow rates and sampling intervals varied from 80-200 mL/min and 1.5-5.5 hours. Flow rates were measured at the start and finish of sample collection using Gilmont Instruments (Barrington, IL) size #11 rotameter-type flow meters (0-250 mL/min). The two rates were averaged and multiplied by the sampling interval to give the total volume of air sampled. All other sampling stations were connected to Model 222-4 low flow sample pumps (SKC-West). Sample pumps were

Table II. Methyl bromide field sampling schedule and number of air samples collected during each sampling interval

Mast	8/12 PM	8/13 AM	8/13 PM	8/14 AM	8/14 PM	8/15 PM	8/16 PM	8/17 AM	8/17 PM	8/18 AM
A				4	4	4	4		4	
B		4	4	4	4	4	4		4	
C		4	4	4	4	4	4		4	
D		4	4	4	4	4	4		4	
E		4	4	4	4	4	4		4	
F		4	4	4	4	4	4		4	
G		4	4	4	4	4	4		4	
H		4	4	4	4	4	4		4	
I				4	4	4	4		4	
J				4	4	4	4		4	
K							4		4	
L							4		4	
T	12	12	12	12	12	12	12	12	12	12
U	12	12	12	12	12	12	12	12	12	12
V	12	12	12	12	12	12	12	12	12	12
W		10	10	10	10	10	10		10	
X		12	12	12	12	12	12		12	
Y					12	12	12		12	

assigned unique identification numbers and were calibrated before use to determine flow correction factors. Pump meter numbers were noted before and after sampling and the calibration correction factors were multiplied by the meter difference to give the milliliter volume of air sampled. After sample collection, sampling trains for all stations were separated and each glass tube was capped and stored on dry ice for shipment to the laboratory, where they were stored at -20°C (~ 3 months) prior to analysis.

Sample Preparation and Analysis. The capped sampling tubes were removed from the freezer and allowed to warm to room temperature. The caps were then removed, the charcoal and glass wool (omitting the polyurethane foam spacers) in each tube were transferred to separate 22 mL glass headspace vials (Perkin-Elmer, Norwalk, CT), 2.5 mL benzyl alcohol (Aldrich Chemical Co., Milwaukee, WI) was added to each vial, and each vial was sealed immediately after solvent addition with a crimped aluminum cap containing a Teflon-coated butyl rubber septum and aluminum star spring (Perkin-Elmer).

Methyl bromide analysis was accomplished using a Perkin-Elmer AutoSystem gas chromatograph coupled to an HS-40 Headspace Sampler. Samples were chromatographed on a 27 m x 0.32 mm (id) PoraPlot Q PLOT column (ChromPak, The Netherlands) and detected using a ^{63}Ni electron-capture detector, both contained in the AutoSystem gas chromatograph. Data were collected and processed on a computer-based integrator using PE Nelson chromatography software.

Each sealed headspace vial was placed into the sample tray of the headspace sampler and the sampler was programmed to thermostat each vial at 60°C for 10 minutes. After the thermostating time, the sampler was also programmed to inject the equilibrated headspace for a pre-set time in the range 0.01-0.2 min, depending upon the amount of methyl bromide in the sample. By setting the PLOT column head pressure at 138 kPa, which gave a column flow of about 3.5 mL/min, an injection time of 0.1 min, for example, resulted in an injected volume of headspace of 350 μL ([0.1 min] x [3.5 mL/min] x [1000 μL/mL]). The PLOT column temperature was maintained at 90°C for 8 min, after which time the column was heated using three temperature ramps before returning to the starting temperature (total analysis time per sample was about 24 min): 10°C/min to 130°C (hold for 2 min), then to 150°C at 10°C/min (hold for 2 min), and then to 210°C at 10°C/min (no hold). This temperature regime is recommended to avoid any cracking of the silica PLOT column due to sudden temperature changes. The following temperatures were also maintained: (1) Electron-capture detector -- 350°C; (2) Sampling needle -- 170°C; and (3) Transfer line -- 170°C.

Samples were quantitated by comparing their instrument responses to those of standard methyl bromide in ethyl acetate spiked to clean charcoal and prepared for analysis using the same method as for the field samples. Standard curves consisted of at least four points, three determinations per point, spanning a range of 2-3 orders of magnitude.

A few surface soil samples, taken at random to a depth of about 1 cm in the adjacent (downwind) strawberry field, were analyzed for methyl bromide by adding about

5 g of well-mixed soil to separate 22 mL headspace vials, adding 2.5 mL benzyl alcohol to each vial, and then sealing each vial and analyzing as before for the air samples.

Results and Discussion

Analytical Methods. Activated charcoal was the adsorbent of choice for the approximately 900 air samples generated in this project because of its ability to interact chemically with methyl bromide to trap it from an air stream. The trapping efficiency of charcoal, determined at a flow of 100 mL/min by either spiking the intake of a charcoal-filled sampling tube with methyl bromide standard in ethyl acetate or by drawing air through a sampling tube connected to a Tedlar bag containing methyl bromide vapor, was 82-85%. While on dry charcoal, methyl bromide appears to be fairly stable, especially at -20°C where methyl bromide can be retained for a prolonged period of time.

The results for the charcoal tubes were expressed as mass methyl bromide per volume of air sampled ($\mu g/m^3$). The secondary tube in the sampling train was used to determine methyl bromide breakthrough. If methyl bromide residues in the secondary tube exceeded 25% of the total residue for both tubes, then the final result would be expressed as a "greater than" number. For most of our field samples, measurable residues resided on the primary tube, and in most of those cases where measurable residues were found on the secondary tubes, levels were only a few percent of the total residue trapped. The practical methyl bromide detection limit for the method was approximately 20 ng/m³ (100 mL/min flow for 4 hrs), which was equivalent to about 0.5 ng methyl bromide spiked to a charcoal tube. Most of the field samples that contained methyl bromide were well above this limit, with some near 100 μg (~4,000 $\mu g/m^3$). Regardless of field sample concentration, chromatograms did not show any interferences near the methyl bromide peak.

Field Design. The treated field was bare soil treated with 67:33 Tricon mix injected to 25 cm, and then tarped with polyethylene -- all in accord with common local practice in the Salinas Valley. The vertical flux mast (Station T) was placed near the center of the field after application and tarping were completed. Wind speed and direction were monitored continuously in the field, and at several of the downwind profile masts. Lateral or remote samplers were collected at a single height toward the expected outward boundaries of the downwind plume. A few other samples were collected at I and J, and at another mast at Y, because the wind direction veered more to this eastward direction during some sampling periods. Most of the downwind sampling was done within a ca 20 ha field of mature strawberry plants located approximately northeast of the treated field -- the predominant wind direction during many of the sampling periods, particularly in the afternoon.

Vertical Flux From Treated Field. For Station T, which was in the center of the treated field, concentration gradients were used in the aerodynamic method *(3,8)* to determine the vertical evaporative flux of methyl bromide out of the treated field. The concentration vs height profile observed over all of the sampling intervals for station T was of the shape expected (See Figure 2 for Station T data from Day 1 and the model-predicted profile shape). Results indicated that the cumulative volatilization loss of methyl bromide over the six day period of the study was about 26% of the applied material, with half of this loss occurring within the first 24 hours (Figure 3). These results are tempered by the fact that sampling was done only during the daylight hours and it is probable that we missed some appreciable flux during the nighttime hours. Even so, these results were similar to those observed for another tarped field situation where the cumulative volatilization loss of methyl bromide was about 22% of the applied material over a five-day period, with half of this loss occurring within the first 24 hours *(3)*. A non-tarped field in this latter study showed 89% loss of applied material in five days.

Integrated Horizontal Flux (IHF). For IHF, the product of concentration and wind speed was integrated over the plume depth *(9, 10)* at Station U, approximately 6 m from the downwind edge of the treated field. This flux term was calculated as a measure of off-site, horizontal mass flow of methyl bromide. The plume depth was determined from an extrapolation of concentration vs height to zero concentration, where the corresponding height would be the plume depth. This type of determination can be readily made as long as concentration decreases with height, as was the case for Station U. The data in Figure 4 were for days 1 and 5, but they typified each day of the study in which flux sampling was conducted.

IHF was compared with vertical flux, determined at Station T using the aerodynamic method, as a useful cross-check for a methodology (flux measurement under open field conditions) which has a notoriously high error associated with it *(3, 8, 10)*. As indicated in Table III, the vertical flux of methyl bromide leaving the field (Station T) was virtually the same as for that passing through a vertical plane downwind from the field (Station U), reinforcing both numbers. The fact that Station U was slightly off-site (~6 m) did not appear to matter, since the depth of the source (~362 m) was so much greater and concentration vs height had an inverse relation (Figure 4).

One could, in theory, calculate a similar IHF term for methyl bromide vapors passing through a hypothetical vertical plane of air at downwind sampling sites further removed from the field. It would be desirable to know the IHF term at increasing distances from the source in order to discern possible methyl bromide sinks (e.g., plants and soil) which, if they are significant, would be reflected in a decrease in IHF with downwind distance. However, the shape of the plume may be such as to preclude plume depth determination, and thus IHF. For example, at Station V the concentration increased with height up to and including the greatest measuring height (2 m) (Figure 5). This may have been due to plume rise, although capture of methyl bromide by plants or soil cannot be ruled out as a possible cause for residue depletion near the surface. Since IHF could not be determined at the stations further removed than Station U, we took a different approach, which involved analysis of downwind surface soil, to indicate whether surface soil was serving as a sink for the methyl bromide vapor plume.

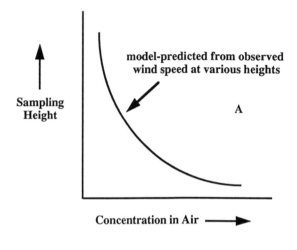

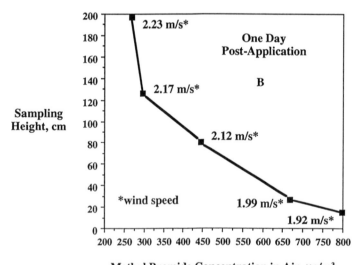

Figure 2. Model predicted vertical pollutant concentration profile for ground-level plume (A) and for measured vertical methyl bromide concentration for Station T (center of treated field) on Day One following treatment (B)

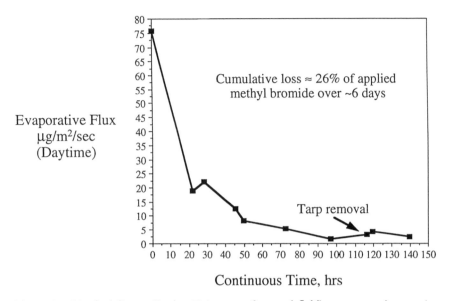

Continuous Time, hrs

Figure 3.   Vertical flux at Station T (center of treated field) versus continuous time post-treatment.

Table III. Comparison of integrated horizontal flux (IHF) determined at Station U (~6 m from downwind edge of the field) with vertical aerodynamic flux determined at Station T (center of field)

Date	Flux, $\mu g/m^2/sec$			
	Aerodynamic[a]	Plume ht, m[a]	IHF[b]	Plume ht, m[b]
8/13/94	22.2	7.1	22.7	15.5
8/14/94	12.4	4.1	13.1	19.7
8/16/94	1.64	4.9	2.30	19.1
8/17/94	4.20	3.1	2.00[c]	15.0

[a] Station T.

[b] Station U.

[c] Data scatter made it difficult to determine concentration and wind speed profiles accurately.

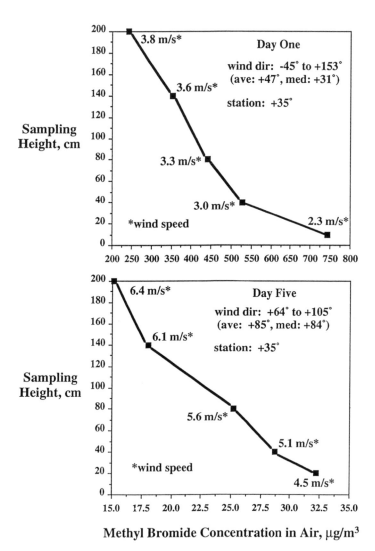

Figure 4.   Concentration vs height for Station U (6 m from downwind edge of treated
field) for Day One (top) and Day Five (bottom) following treatment

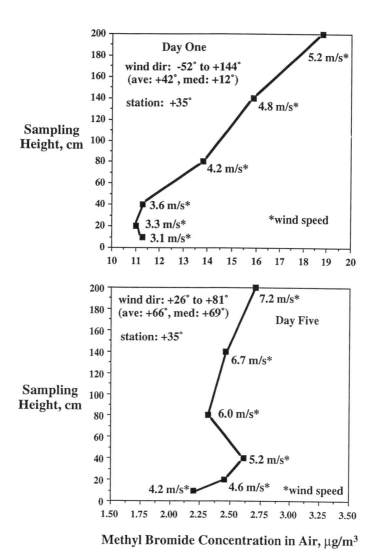

Figure 5. Concentration vs height for Station V (100 m from downwind edge of treated field) for Day One (top) and Day Five (bottom) following treatment

With regard to the possibility of methyl bromide adsorption to downwind soil, we analyzed a few soil surface samples (to approximately 1 cm depth) collected at random from the strawberry field downwind from the treated field. Our analyses showed the presence of methyl bromide, confirmed by mass spectrometry, at levels in the range of 15-150 μg/m^2. These results were compared to those of an earlier study (11) where surface concentrations at equilibrium were about 2.5 g/m^2 for dry sand, 6.5 g/m^2 for dry clay soil, and 10.5 g/m^2 for dry peat at a methyl bromide partial pressure of about 6.68 torr. Assuming Langmuir adsorption, where the amount of adsorbate on the adsorbent is proportional to the adsorbate partial pressure, and taking a mid-point between sand and clay (i.e., 4.5 g/m^2) for a sandy clay with an organic matter content of about 1-3%, we used a simple proportionation between our field data and the data of the cited study to calculate a concentration range for methyl bromide in field air. Based on our dry soil residue results, the air concentrations responsible for these observed residues fell in the range 100-1,000 μg/m^3. Observed air concentrations for the first few sampling periods of the study fell in this range. Unfortunately, some soil samples taken at different distances downwind were inadvertently mixed so that we are only able to report a range of preliminary residue levels. However, these field results were supported by preliminary results from the laboratory that indicated that for a methyl bromide vapor density of about 248 μg/m^3 above Salinas Valley soil in sealed vials, soil surface concentrations fell in the range 8-12 μg/m^2. Furthermore, vapor-soil equilibrium was established in about 2-4 minutes at 40°C. Adsorption of methyl bromide from a plume by soil was studied in more detail in a different field experiment to be reported separately.

Downwind Concentration vs Distance. The air concentration of vapors as a function of downwind distance is expected to decline regularly so long as the wind and terrain are relatively regular. The concentration-distance profile is useful information for estimating exposures, and setting buffer zones. Using a single sampling height (2 m), we determined these profiles for Stations U, V, W, and X . The data were then plotted as air concentration vs downwind distance (Figure 6). The plots were similar in slope even though the wind direction was not always optimally toward the sampler, and, of course, the concentrations varied greatly depending on the day of sampling. Furthermore, for many of the sampling periods, air concentration showed a logarithmic decline with distance from the source. Similar declines of concentration vs distance from the field have been reported for other pesticides (12).

Modeled vs Measured Downwind Concentrations. The Industrial Source Complex Short Term II (ISC-STII) Model is widely used for estimating atmospheric concentrations of a chemical downwind from a source (13). This model is based upon the Gaussian plume dispersion equation, providing a proportional relationship between flux from the source (field) and downwind air. It can also be used to back-calculate a flux term when the downwind concentrations are known, for comparison with measured flux (14).

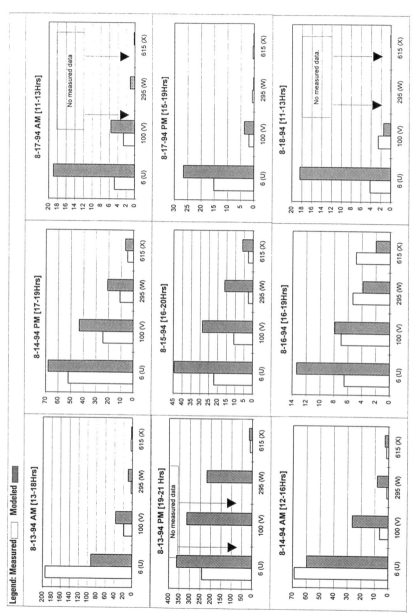

Figure 6. Graphs of methyl bromide measured and modeled concentration vs distance at 2 meter heights

Table-IV: Vertical Flux Values for Methyl Bromide Corresponding to ISC-STII Modeled Intervals		
Day	Time (hh:mm)	Vertical Flux (μg/m2/s)
8/12/94	14:20 - 19:48	75.8
8/13/94	12:09 - 16:56	18.9
	18:31 - 20:15	22.2
8/14/94	11:19 - 14:45	12.4
"	16:09 - 19:00	8.2
8/15/94	14:48 - 18:51	5.24
8/16/94	15:02 - 18:46	1.64
8/17/94	10:39 - 12:48	3.3
"	14:01 - 18:27	4.21
8/18/94	10:11 - 12:06	2.52

The ISC-STII model estimates hourly concentration at downwind distances assuming a straight line transport of pollutants from the source by the prevailing wind under steady weather conditions. It is based on the equation for a finite crosswind line source. The model requires that individual area sources have the same north-south and east-west dimensions. When an irregular area source is encountered its effect can be simulated by dividing the area source into multiple squares that approximate the geometry of the irregular area source. The size of the individual area source could vary but the only requirement is that each area source must be a square. The model requires that, if the source receptor separation is less than the length of the side of the area source, then the area source should be subdivided into smaller area sources because the finite line segment algorithm of the model does not adequately represent the source-receptor geometry.

The inputs for the ISC-STII are primarily weather data such as wind speed, wind vector, surface air temperature, stability class and urban/rural mixing height. All the required weather data were obtained by a 2m tall weather station set up at the center of the field. Mixing heights for Salinas were based on the radiosonde measurements from the National Weather Station at Oakland, CA., for the same time period, and they were approximated to be 350m. Stability class estimates were made from wind speed, solar insolation and state of sky (*15*). A key input into the model is the source strength which was obtained from the vertical flux measurements made at the center of the field (Table IV). The model outputs are downwind methyl bromide concentration averages for the period indicated in Table IV at a uniform mast height of 2m.

Since the ISC-STII model requires only square sources be used for estimating downwind concentrations, the 15.6 ha field was approximated to a square of 396 m x 396 m and this square was further divided into 169 sub-squares of 30.5 m x 30.5 m. All of the 169 sources were referenced with respect to a coordinate axis whose origin was fixed at the north-east corner of the field. A similar receptor grid was constructed with reference to the same origin. As the field was oriented at an angle of +13° with regard to N, for ease of modeling it was rotated in the counter-clockwise direction by 13° and brought in alignment with the cardinal directions. Accordingly, the wind vector was also rotated by 13° in the counter-clockwise direction. The straight line on which the field sites U, V, W and X were aligned was inclined at an angle of +22° with respect to N (after 13° counter-clockwise rotation; 35° without rotation) and their distances from the center of the field are given in Table I.

Wind directions (the direction from which the wind is blowing with respect to north) were transformed to wind vectors (the direction into which the wind is blowing with respect to north). The ISC-STII model is constructed to accept only hourly weather data and therefore the weather data collected at 5 minute intervals were vectorially averaged. The vectorial averages of wind velocities and directions and means of temperature were used as input to the model.

The model yielded concentrations downwind of the treated field which were then plotted as concentration contours or isopleths using the graphical software package SURFER Version 5.01. Figure 7 shows a typical isopleth. Table V compares the

Table V : Modeled vs Measured Downwind Concentrations of Methyl Bromide based on the ISC-STII Model

Session	Station (distance from centre of field)	Modeled (μg/m^3)	Measured (μg/m^3)	Measured Values vs Modeled Values	Modeled values > measured values (%)
8/13/94	U (186m)	92.02	194.22	High[1]	47
Session-1	V (273m)	37.26	18.82	Low	198
13-18 (Hrs)	W (453m)	9.77	2.97	Low	329
	X (792m)	2.41	3.57	High	68
8/13/94	U	362.25	241.83	Low	150
Session-2	V	314.43	No Data	No Comparison	No Comparison
19-21 (Hrs)	W	219.01	No Data	No Comparison	No Comparison
	X	15.19	10.7	Low	142
8/14/94	U	59.74	68.76	High[1]	87
Session-1	V	26.55	6.38	Low	416
12-16 (Hrs)	W	8.27	1.16	Low	713
	X	2.75	1.56	Low	176
8/14/94	U	68.09	52.08	Low	131
Session-2	V	43.42	24.43	Low	178
17-19 (Hrs)	W	20.92	10.99	Low	190
	X	6.63	4.83	Low	137
8/15/94	U	44.48	22.03	Low	202
16-20 (Hrs)	V	28.67	10.83	Low	265
	W	16.11	2.58	Low	624
	X	6.04	2.68	Low	225
8/16/94	U	13.37	6.54	Low	204
16-19 (Hrs)	V	7.9	7	Low	113
	W	3.87	5.32	High	73
	X	2.02	4.85	High	42
8/17/94	U	18.93	4.8	Low	394
Session-2	V	5.64	2.7	Low	209
11-13 (Hrs)	W	1.1	No Data	No Comparison	No Comparison
	X	0.21	No Data	No Comparison	No Comparison
8/17/94	U	26.6	15.13	Low	176
Session-2	V	3.67	1.89	Low	194
15-19 (Hrs)	W	0.2	0.68	High	29
	X	0.19	0.45	High	42
8/18/94	U	18.4	4.15	Low	443
11-13 (Hrs)	V	1.48	2.56	High	58
	W	0.03	No Data	No Comparison	No Comparison
	X	0.07	No Data	No Comparison	No Comparison
% "Lows"		73		%average =	209
% "Highs"		20			
		7		1: Dist. of sampler from edge of field (5m) < length of side of area source (33m)	
# of Lows		22			
# of Highs		6		% > measured = (modeled/measured)*100	
# of "na"		6			

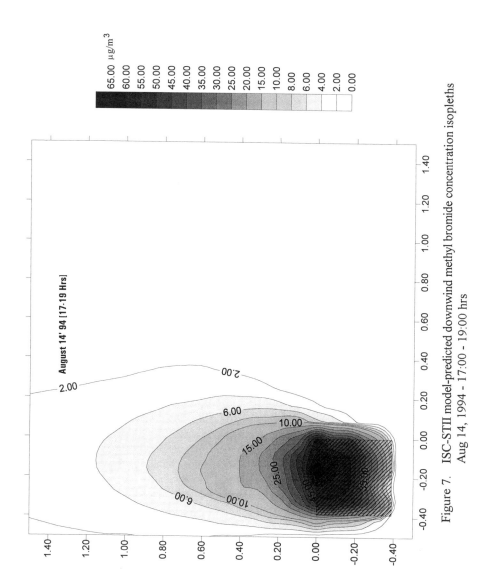

Figure 7. ISC-STII model-predicted downwind methyl bromide concentration isopleths Aug 14, 1994 - 17:00 - 19:00 hrs

modeled results with the measured results for sites U, V, W and X. Figure 6 is a graphical depiction of this comparison. Thirty-six modeled results were available for comparison with thirty measured concentrations. As expected from results of a prior study (*14*), the measured concentrations were lower than the modeled for ~73% of available data. The modeled results on average were twice the measured concentrations with a few extreme cases where the modeled values were higher by a factor of 6. These greater deviations were observed for sites which were not in line with the prevailing wind during the particular sampling session.

Station U was located at 6m from the downwind edge of the field, a distance that is less than the length of one side of the square source (33m), under which situations the model is not expected to provide accurate results. Thus, for modeled values to be lower than measured concentrations at Station U during Session-1 on 8/13/94, and 8/14/94, is not unexpected. Six (~20%) of the measured values were higher than the modeled concentrations. For the most part, these higher values occurred at the more remote stations which were most affected by ambient concentrations of methyl bromide from treatments carried out at other fields in the general vicinity of this experimental field. Two stations located upwind of the experimental field (A and B) confirmed the presence of ambient methyl bromide intrusion, which was of a similar magnitude to that recorded at Stations W and X.

Summary and Conclusions

Two of the many questions associated with the contribution of agricultural use of methyl bromide as a soil fumigant to stratospheric ozone depletion were addressed in an extensive field application-sampling regime in the Salinas Valley of California. This experiment provided an estimate of vertical flux from the tarped field of 26% of the amount applied over a 6-day period following application. This estimate, based upon flux measurements collected during daytime intervals, is in good agreement with a prior reported flux loss of 22% of material in a tarped field situation in the Salinas Valley (*3*). Other investigators have reported values as high as 87% of the applied material from fumigated and tarped fields, much higher than the values we observed in these two trials. The present experiment included an internal check on our flux methodology. Horizontal flux measurements conducted 6 m outside the downwind edge of the treated field provided flux estimates in good agreement with those made at the center of the field using the aerodynamic gradient method. We recommend that the horizontal flux be calculated whenever possible as a check, because of the large error associated with flux measurements in large open areas.

A second question deals with the possible operation of terrestrial sinks to remove methyl bromide vapors from the air. Prior assumptions, by U.S. EPA and others, are that all of the methyl bromide which escapes from treated fields can potentially diffuse to the stratosphere, where it may photochemically degrade to bromine atoms which, in turn, destroy ozone. The ozone depleting potential of bromine is about 40 times that of chlorine. What is not known is whether sinks exist for removing methyl bromide before diffusion to the stratosphere occurs. Reactivity with hydroxyl radical represents one possibility, but the reaction rate in the atmosphere is reported to be too slow to significantly affect destruction prior to diffusion to the stratosphere *(6)*. Another possibility, exchange with the oceans, is under intense study by others, and it appears that this may represent both a source and a sink, but a net sink, at least for some ocean locations *(16)*. With regard to terrestrial sinks, not much is known. Absorption of methyl bromide by plant foliage is one possibility, exchange with surface waters is another, and adsorption to soil is a third. If any/all of these processes are partly or completely irreversible, then they would represent net sinks for airborne methyl bromide. Shorter et al *(17)* recently reported the rapid and irreversible removal of atmospheric methyl bromide by soils. In the present study we examined three methods for obtaining data on the operation of near-source soil and vegetative sinks. One involved measuring the methyl bromide plume behavior, as concentration vs height profiles, downwind of the source. A lower than expected concentration near the soil/vegetation surface could indicate removal, or capture, of methyl bromide by the surface *(7)*. We obtained experimental data showing a depletion of concentration near the surface, but we were unable to differentiate between a surface capture mechanism and simple plume lift-off from the surface caused by local meteorological conditions.

A second method involved analysis of soil samples from the downwind surface. We collected a few composited samples and were able to measure intact methyl bromide. This shows that adsorption is occurring, supported by laboratory measurements in sealed vials, but does not address the question of reversibility -- that is, whether adsorbed methyl bromide is transiently bound, and then eventually released, or whether it degrades following adsorption. This needs to be addressed more completely, with soil and also with vegetation.

A third method involved comparison of measured downwind air concentrations with those predicted by a Gaussian plume dispersion model which has no sink terms in it. If the measured values were significantly lower than those predicted, and if the differential increased with downwind distance, the operation of sinks could be inferred. Our measured downwind values were, in fact, generally lower than those predicted by the ISC-STII model. But there was much uncertainty in both the measured and modeled values and in the consistency of the wind patterns, and the differential did not show a regular and consistent increase with downwind distance. Thus, although this approach may be quite viable for estimating sinks in a large, uniform airshed, it may be too blunt to be definitive for relatively short fetches such as were employed here. On a positive note, the ISC-STII model shows much promise for estimating downwind concentrations of methyl bromide, based upon the results from this study and those of others for Telone-II *(18)*. The ISC-STII model can thus be very useful for estimating exposures and setting buffer zones for the use of fumigants.

In conclusion, we examined approaches for characterizing the downwind behavior and potential loss of methyl bromide vapors. We found an indication that the downwind soil surface may be a sink for methyl bromide. With more refinement, these methods are of potential use in examining downwind sinks, and downwind exposures, for methyl bromide and for other airborne contaminants emitted from large field sources.

Acknowledgments

Tri-Cal Corporation (Dr. Tom Duafala, Mr. Kirk Fowler, Mr. Matt Gillis, Mr. James Saldaño) arranged for the field application and helped with sampling and modeling. Salinas Berry Farms (Mr. Ken Mukai) helped with logistics, including arranging for unimpeded access to sampling locations. Mr. Robert Roach of the Monterey County Agricultural Commissioners Office provided helpful methyl bromide use information. This project was supported, in part, by funds provided by the National Pesticide Impact Assessment Program (Drs. Nancy Ragsdale and Willis Wheeler) and the Western Region Pesticide Impact Assessment Program (Mr. Rick Melnicoe). We also acknowledge assistance in field sampling provided by Dr. David Crohn, UC Riverside, and by Mr. Chad Wujcik and Mr. Ken Seiber of the University of Nevada, Reno.

Literature Cited

1. Andersen, S. O. and Lee-Bapty, S. *Methyl Bromide Interim Technology and Economic Assessment*, **1992**, United Nations Environment Program.

2. Yagi, K., Williams, J., Wang, N. Y. and Cicerone, R. J. *Proc. Natl. Acad. Sci. USA.*, **1993**, *90*, 8420.

3. Majewski, M. S., McChesney, M. M., Woodrow, J. E., Pruger, J. H. and Seiber, J. N. *J. Environ. Qual.*, **1995**, *24*, 742.

4. Mellouki, A., Talukdar, R. K. and Howard, C. J. *Methyl Bromide Global Coalition State of the Science Workshop*, **1993**.

5. UNEP. *United Nations Environment Program (UNEP).* Montreal Protocol Synthesis Report of the Methyl Bromide Interim Scientific Assessment and Methyl Bromide Interim Technology and Economic Assessment, **1995**.

6. Grosjean, D. J. *Air Waste Manage. Assoc.*, **1991**, *441*, 56.

7. Parmele, L.H., E.R. Lemon and A.W. Taylor. *Water Air Soil Pollut.* **1972**, *1*, 433.

8. Majewski, M. S., Glotfelty, D. E., Paw U, K. T. and Seiber, J. N. *Environ. Sci. Technol.*, **1990**, *24*, 1490.

9. Denmead, O.T. In *Gaseous Loss of N$_2$ From Plant-Soil Systems*, Francy, J.R.; Simpson, J.R. (Eds), Martinus Hoff/Dr. Junk: The Hague, **1983**, 133.

10. Glotfelty, D.E., Shomberg, C.J., McChesney, M.M., Sagebiel, J.C. and Seiber, J.N., *Chemosphere*, **1990**, *21*, 1313.

11. Chisolm, R. D. and Koblitsky, L. *J. Econ. Entomol.*, **1943**, *36*, 549.

12. Seiber, J.N. and Woodrow, J.E. *Arch. Environ. Contamin. Toxicol.*, **1981**, *10*, 133.

13. Wagner, C. P. In *Industrial Source Complex (ISC) Dispersion Model Users Guide*, Office of Air Quality Planning and Standards, US Environmental Protection Agency, Washington, DC, 1987, Vol. 1.

14. Ross, L. J., Johnson, B., Kim, K. D. and Hsu, J. *Environmental Hazards Assessment Program*, Department of Pesticide Regulation, Sacramento, CA, 1995, Vol. EH-95-03

15. Pasquill, F.A. and Smith, F.B. In *Atmospheric Diffusion, 3rd Edition*, Ellis Horwood, **1983**, 336.

16. Lobert, J. N., Butler, J. H., Montzka, s. A., Geller, L. S., Myers, R. C. and Elkins, J. W. *Science*, **1995**, *267*, 1002.

17. Shorter, J.H., Kolb, C.E., Crill, P.M., Kerwin, R.A., Talbot, R.W., Hines, M.E. and Harriss, R.C. *Nature*, **1995**, *377*, 717.

18. Fontaine, D.D. and Weinberg, J.T. Paper presented at the 210th National Meeting of the American Chemical Society (AGRO 92), Chicago, IL, Aug 20-24.

Chapter 14

Off-Site Air Monitoring Following Methyl Bromide Chamber and Warehouse Fumigations and Evaluation of the Industrial Source Complex–Short Term 3 Air Dispersion Model

T. A. Barry, R. Segawa, P. Wofford, and C. Ganapathy

Environmental Monitoring and Pest Management, Department of Pesticide Regulation, California Environmental Protection Agency, 1020 N Street, Sacramento, CA 95814–5624

The Department of Pesticide Regulation's preliminary risk characterization of methyl bromide indicated an inadequate margin of safety for several exposure scenarios. Characterization of the air concentrations associated with common methyl bromide use patterns was necessary to determine specific scenarios that result in an unacceptable margin of safety. Field monitoring data were used in conjunction with the Industrial Source Complex - Short Term (ISCST3) air dispersion model to characterize air concentrations associated with various types of methyl bromide applications. Chamber and warehouse fumigations were monitored and modeled. For each fumigation the emission rates, chamber or warehouse specifications and on-site meteorological data were input into the ISCST3 model. Linear regression analysis was used to compare the model-predicted concentrations to measured air concentrations. The concentrations predicted by the ISCST3 model reflect both the pattern and magnitude of the measured concentrations.

Methyl bromide is one of the most widely used pesticides, with 6.7 million kilograms applied in California in 1993 (1). The largest quantity of methyl bromide is used for soil fumigation in agricultural fields. However, the largest number of applications occur for other uses, often in populated areas where residential structures are in close proximity. These uses include the fumigation of harvested commodities in chambers, truck trailers, sea containers, storage silos, warehouses, mills, and under tarpaulins, and for general pest control in food processing plants, houses, and apartments. Other minor uses include fumigation of soil in greenhouses, potting soil, turf areas for pre-planting, and packaging material.

    The Department of Pesticide Regulation's (DPR) preliminary risk characterization of methyl bromide indicated that an inadequate margin of safety existed for several

0097–6156/96/0652–0178$15.00/0

exposure scenarios (2). As a consequence, DPR developed permit conditions for uses of methyl bromide in the state of California (3,4). These use permits include buffer zones surrounding applications. Calculation of buffer zone size required the characterization of air concentrations associated with each type of use.

There are over 100 different uses of methyl bromide in California. Use of monitoring alone was not possible because it would require a very extensive monitoring program to provide sufficient spatial and temporal coverage. As a result, it was necessary to also use an air dispersion model. Air dispersion models are powerful tools for calculating health protective buffer zones and to screen candidate mitigation measures (5). However, in order to use an air dispersion model for these purposes, it was necessary to gather a foundation set of monitoring data for the most common use scenarios. DPR conducted field monitoring to assemble this foundation data set. The objectives of these studies were three-fold:

1) Characterize methyl bromide air concentrations associated with commodity chamber fumigations during the aeration phase
2) Characterize methyl bromide air concentrations associated with warehouse space fumigations during both the treatment and the aeration phases
3) Develop the appropriate model inputs to insure that the Industrial Source Complex - Short Term (ISCST3) air simulation model (6) reliably predicts downwind air concentrations resulting from commodity chamber and warehouse space fumigations.

Commodity Fumigation -

Commodity fumigation is typically conducted in a fumigation chamber. Produce including plums, peaches, cherries, nuts and grapes are fumigated to improve storage life or to satisfy quarantine requirements. DPR monitored a total of five commodity fumigation chambers during 1992. Monitoring and modeling results from one chamber fumigation are presented in this paper. This study monitored the air concentrations associated with fumigation of walnuts.

Chamber and Fumigation Characteristics. This chamber is situated within the southeast corner of a larger, main building. The main building measured 81 m by 68 m and was 6.1 m tall. The chamber had a volume of 397 m³, a chamber height of 6.1 m and an exhaust stack height of 7.9 m from ground level. Although the exhaust fan capacity was 0.38 m³/s, the stack had a 90° elbow at the top. This resulted in a vertical exit velocity of nearly zero. A total mass of 13.6 kg of methyl bromide was injected into the sealed chamber to achieve the target application rate of 8840 ppm. The treatment period duration was 17 hours.

Air Sampling. Air sampling was conducted only during the aeration phase of this fumigation. Air concentrations were measured in the chamber stack as well as at one upwind and several downwind locations. Downwind samplers were deployed at seven locations at distances of 52 to 250 m from the stack. Air concentrations were measured by two different methods: 1) the initial high concentrations in the stack were measured on a real time basis using a thermal conductivity detector, and 2) Charcoal tubes were used to sample the stack after concentrations were not detectable with the thermal conductivity detector (below 500 ppm), and for all field samplers downwind and upwind. In this method, two charcoal tubes (primary and back-up), connected end-to-end, were attached to an air pump. Methyl bromide was trapped on the charcoal as air was drawn through the tubes by the air pump. Laboratory analysis of the charcoal tube samples was conducted by the California Department of Food and Agriculture's Chemistry Laboratory Services. The charcoal from each primary and backup tube was extracted separately with carbon disulfide. The resulting extract was then analyzed by a gas chromatograph equipped with an electron capture detector. Laboratory spiked and blank samples were also analyzed for quality control. The air flow rate of each sampler was adjusted with the duration of the sample so that the volume of air remained constant at approximately 11 liters. Therefore, the detection limit for all samples was 0.2 µg/sample, equivalent to approximately 0.005 ppm.

The field samples were taken at intervals of 30 minutes. Two 30-minute samples were averaged to obtain a one hour average concentration for comparison with modeling results. Stack air flow rates and on-site meteorological data were also measured.

ISCST3 Dispersion Modeling. The ISCST3 dispersion model was used to simulate air concentrations associated with the chamber stack releases of methyl bromide. ISCST3 is a Gaussian plume dispersion model that uses stack emission rates, chamber characteristics and meteorological data to predict air concentrations. One-hour average air concentrations were simulated for discrete receptors corresponding to the air samplers in the field. The effects of the air flow over and around the main building within which the chamber was located were included in the modeling. The main building was large so there was significant potential for building effects to act on the released plume. Since the stack on the chamber at this site has a 90° elbow, the effective vertical exit velocity was set to 0.001 m/s. This exit velocity was based on EPA guidelines for obstructed stacks (7) . Stack air concentrations and the measured air flow rate were used to calculate the mass of methyl bromide emitted in the first 60 minutes. This mass was used to calculate a flux rate for the stack in units of grams per second (g/s) for 60 minutes.

Statistical analysis matched each air sampler with the appropriate model receptor. Linear regression analysis, fitting the modeled concentration as a function of the measured concentration, was used to assess the ability of the model to reproduce the magnitude and spatial pattern of air concentrations observed in the field. Student's t-tests were used to determine whether the regression analysis showed reliable correspondence between the measured and modeled results. Student's t-test results are reported as follows:(t=calculated value, df = degrees of freedom, p = p-value).

Results. Initial stack air concentrations and emission rates were very high, but declined rapidly over time. The initial stack concentration was 6400 ppm, 76% of the application rate. A total of 8.6 kg were released within the first 60 minutes of the aeration period. The total mass released during the 2-hour aeration period was 9 kg. The remainder of the 13.6 kg applied was either absorbed into the commodity or leaked from the chamber during the holding period. These results are characteristic of fumigation chamber fumigation/aeration events.

Figure 1 shows the location of the air samplers relative to the stack and the 1-hour average concentration measured during the first hour of aeration at each sampler. The highest concentration measured was 0.115 ppm 52 meters downwind of the stack. Since the mass released during the aeration of this chamber was very small, all concentrations measured were well below the target 1-hour Time Weighted Average (TWA) concentration of 5.0 ppm (2). However, it should be noted that methyl bromide was detectable 250 m from the stack with a one-hour concentration of 0.031 ppm. The one-hour mean wind direction was from 288° and the mean wind speed was 3.8 m/s. The time of day, wind speed, and standard deviation of horizontal wind direction indicated that the atmospheric stability class during the aeration was C, slightly unstable. Concentrations during the second hour of the aeration were very low or not detectable.

The modeled and the field measured concentrations showed a correlation coefficient of 0.72 (Figure 2). The regression of modeled versus measured concentrations was:

$$\text{modeled} = 0.0051 + 1.49 * \text{measured} \quad R^2 = 0.53$$

The regression coefficient was significantly different from zero (t=2.35, df=4, p=0.06) but was not significantly different from 1.0 (t=0.77, df=4, p=0.25). In addition, the intercept was not significantly different from 0 (t=0.13, df=4, p=0.90). This regression equation indicates that the model reproduced both general pattern and magnitude of the measured concentrations. Similar results were obtained at the remaining four chamber fumigation sites.

Warehouse Space Fumigations -

Several studies were conducted to sample the air concentrations associated with a methyl bromide space fumigation of warehouses and food processing facilities. These types of fumigations are typically applied to rid the facilities and equipment of pests. The space fumigation studies were conducted to characterize the leakage of methyl bromide from the warehouse during treatment, to characterize the air concentrations associated with the aeration of the warehouses (aeration may be by forced air or passive aeration), and to compare the measured concentrations with those predicted by the air dispersion model estimates. This type of application is of concern because large amounts of methyl bromide are used and the warehouses leak an unknown fraction of the applied methyl bromide. Results from one study are presented in this paper.

Warehouse and Fumigation Characteristics. The warehouse was a 67 m by 61 m concrete building with a volume of 41060 m³. The majority or the warehouse was 8.5 m tall. On the northeast corner of the building was a second tier that had a height of 14.6 m from ground level. Three exhaust fans were located on the roof, each with a release height 1.8 m above the roof line and a vertical exit velocity of 4.7 m/s. To achieve the target application rate of 6200 ppm, a total of 989 kg of methyl bromide was injected into the sealed warehouse. The treatment period duration was 24 hours.

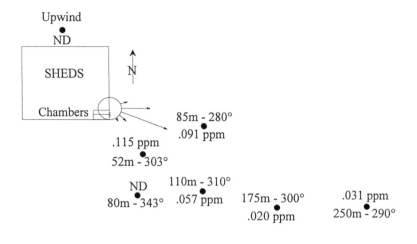

Figure 1. Chamber fumigation site diagram indicating the positions of the air samplers relative to the chamber and the average air concentrations (ppm) measured in the first hour of aeration.

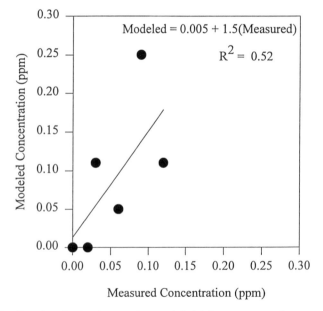

Figure 2. Chamber fumigation results, modeled 1-hour average air concentrations (ppm) as a function of measured 1-hour average air concentration (ppm).

Air Sampling. Air sampling was conducted during both the treatment and the aeration phases of the fumigation. Air concentrations inside the warehouse were measured by a thermal conductivity detector several times during the 24-hour treatment period. Fifteen air samplers were placed around the warehouse in two circles at a distance of 9 m and 30 m from the warehouse. Four consecutive six-hour samples were collected using charcoal tubes connected to air samplers at each location during the treatment period. (See the commodity chamber fumigation section for a description of the charcoal tube sampling and analysis.) Wind speed and direction, temperature and humidity during the treatment period were recorded.

The methyl bromide concentration inside the warehouse was measured by a thermal conductivity detector immediately prior to commencement of aeration. Concentrations during aeration were also measured with charcoal tubes in one of the stacks. Four consecutive 15-minute samples were collected with charcoal tube air samplers at six downwind locations during the first hour of aeration, three samplers were placed at 9 m from the warehouse and three were placed at 30 m from the warehouse. Wind speed and direction, temperature and humidity during aeration were recorded.

ISCST3 Dispersion Modeling. The ISCST3 dispersion model was used to simulate both the treatment period leakage and the release of methyl bromide from the exhaust stacks during aeration. For the treatment period simulation, the warehouse was represented as four area sources at different heights. The first area source, representing leakage from doors and vents near the ground level, was placed at a height of 0.9 m from the ground. The second area source, representing leakage from upper windows and seams was placed at a height of 4.25 m. The third area source, representing leakage from roof vents on the first building tier, was placed at a height of 8.5 m. The fourth area source, representing leakage from the roof vents on the second tier, was placed at a height of 14 m. The three sources representing the first tier had the same area as the warehouse, 4096 m^2. The fourth source representing the second tier had an area of 930 m^2.

The thermal conductivity detector measurements taken inside the warehouse were used to calculate a total mass lost during the 24-hour treatment period. This total mass loss was proportionally divided between the four area sources. A 24-hour average flux rate was calculated based on the mass loss, the area of the sources and the time period. The 24-hour TWA concentration was simulated using 24 hours of one-hour average wind speed and direction calculated from the meteorological data collected on-site.

The aeration period was modeled using three point sources, each representing one of the exhaust stacks. The effects of the air flow over and around the warehouse on the movement of the methyl bromide plumes were included in the modeling. The exhaust stack openings were only 1.8 m above the roof line. Therefore, significant building effects were expected. The one-hour average flux rate for each exhaust stack was calculated based on the stack concentration measurements, the air flow rate and the time period. The one hour average wind speed and direction was calculated from the meteorological data collected on-site.

Results. During the treatment period, measurements of air concentrations inside the processing plant indicate that at least 59% of the applied methyl bromide was retained

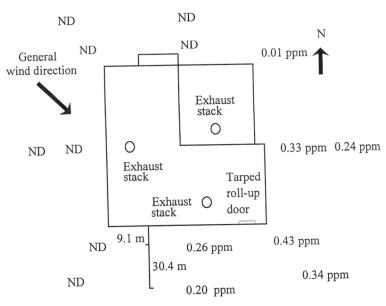

Figure 3. Warehouse space fumigation site diagram indicating the positions of air samplers relative to the building and the average air concentrations (ppm) measured during the 24 hour treatment period.

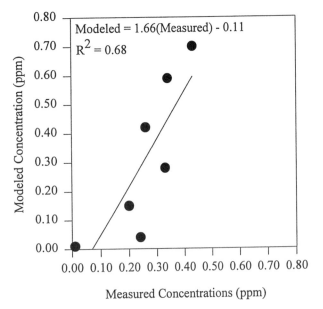

Figure 4. Warehouse space fumigation results, modeled 24-hour average air concentration (ppm) as a function of measured 24-hour average air concentration (ppm).

within the building during the 24-hour treatment period. Consequently, up to 405 kg, or 41%, of the applied methyl bromide leaked from the building. This leakage caused ambient concentrations to exceed the 24-hour TWA target concentration of 0.210 ppm (2) at several receptors (Figure 3). The highest 24-hour TWA concentration was 0.43 ppm at the sampler located 9 m to the southeast. The 0.210 ppm 24-hour TWA target concentration was also exceeded at 30 m from the building at both the east and southeast samplers.

Leakage from a building is the summation of concentrated sources leaking from doors, windows and seams. An initial simulation run using a volume source representation of the warehouse showed poor correspondence with the field data. Concentrations decreased too rapidly both as distance from the warehouse increased and in the cross-wind direction from the center of the plume. A regression of 24-hr TWA observed concentrations compared to the simulated concentrations did not show a significant correspondence. Therefore, the representation described in the methods section using four area sources at different heights was investigated. The modeled and field measured concentrations showed a correlation coefficient of .82 (Figure 4). The regression function of modeled versus measured concentrations was:

Modeled = 1.60(measured) - 0.11 $R^2 = 0.68$

The regression coefficient was significantly different from zero (t=3.24, df=5, p=0.023) and was not significantly different from 1.0 (t=1.29, df=5, p=0.25) and the intercept was not different from zero (t= -0.79, df=5, p=0.47). In addition, the model agreed within a factor of two with the measured concentrations at 6 of the 7 receptors showing concentrations above the detection limit. At all samplers showing non-detects, the model also predicted non-detects. The ISCST3 model replicated the general magnitude and pattern of concentrations measured at the site. Similar results have been obtained at other space fumigations.

Based on initial measurements inside the building, stack concentrations during the first 60 minutes of aeration and the flow rate of the exhaust fans, it is estimated a total of 426 kg of methyl bromide were emitted in the first hour of aeration. A one-hour flux rate of 40 g/s at each exhaust stack was calculated from these estimates. The majority of the mass of methyl bromide released during aeration occurs during the first hour, however, exhaust stack concentrations exceeded 1000 ppm at the end of the 60 minute monitoring period. The warehouse is large and, even though it was mostly empty, more that one hour is required to reduce air concentrations to levels that would permit reentry. The total aeration time for this building fumigation was 24 hours.

Downwind air concentrations during the aeration indicate that the building had a significant influence on the behavior of the exhaust plumes. The stacks are short relative to the building height so the exhaust plumes were entrained in the air flow around the warehouse. High concentrations were detected directly downwind at samplers both at 9 m and 30 m from the warehouse. One hour concentrations of 6.44 ppm and 3.24 ppm were measured at 9 m (Figure 5). One hour concentrations of 3.36 ppm and 5.99 ppm were measured at 30 m. The target levels for the 1-hour (5 ppm) and 24 hour exposure (0.210 ppm) were exceeded at both sampling distances. If the stacks were

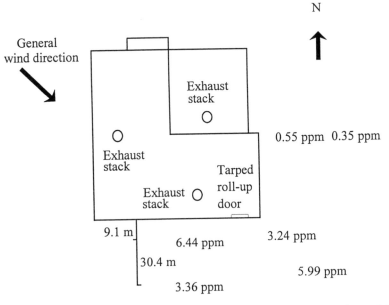

Figure 5. Warehouse space fumigation site diagram indicating the positions of the air samplers relative to the building and the average air concentrations (ppm) measured during the first hour of aeration.

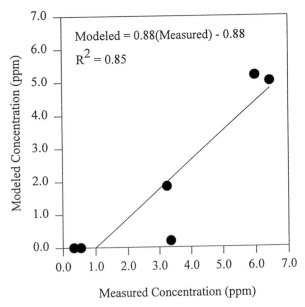

Figure 6. Warehouse space fumigation results, modeled 1-hour average air concentration (ppm) as a function of measured 1-hour average air concentration (ppm).

properly designed for aeration of toxic gases, the methyl bromide released would be carried away from the warehouse. The plumes would not reach ground level for some distance downwind after concentrations had fallen below levels of concern. Combinations of worst case weather conditions (low wind speed and F stability class) and various stack characteristics (stack height, exit velocity, and diameter) can be used to arrive at stack requirements that ensure health levels are not exceeded at any point downwind. These combinations were used as the basis for the stack requirements in the methyl bromide permit conditions (3). Since the stack concentrations exceeded 1000 ppm at the end of the one hour sampling period, it can be expected that high concentrations would continue to be measured downwind during the second hour of aeration.

The ISCST3 model was used to simulate the aeration phase as three point sources, one for each exhaust stack. The effects of the warehouse upon the plume behavior were included in the simulation. Both tiers of the building were used to calculate the potential building effects. Each stack has a one hour flux rate of 40 g/s as calculated earlier. The exit velocity was set to 4.7 m/s. The on-site one-hour average wind speed was 4.1 m/s and the one-hour average horizontal wind direction was from $325°$. The stability class was set to D (neutral) based on the standard deviation of the horizontal wind direction and the presence of cloud cover. The air samplers in the field were represented as discrete receptors. The correlation coefficient of 0.92 indicates that the modeled concentrations agreed very well with the measured concentrations (Figure 6). The regression of modeled versus measured concentrations is:

Modeled = 0.88(measured) - 0.88 $R^2 = 0.85$

The slope was significantly different from zero (t=4.66, df=4, p=0.01) but was not significantly different from 1.0 (t=0.63, df=4, p=0.28) and the intercept was not significantly different from zero (t= -1.14, df=4, p=0.32). Three of the four model receptors agreed within a factor of two with the measured values at the samplers that were directly downwind. The remaining two samplers and model receptors were located on the northeast side of the building, not directly downwind. The modeling results indicated that these locations were highly influenced by the taller, second tier of the building. The ISCST3 model replicated the general magnitude and pattern of concentrations measured at the site.

Conclusions

The commodity chamber fumigation study presented in this paper demonstrates that stack flux rates for this type of fumigation are very high during the first hour of aeration. The majority of the mass of methyl bromide release from these facilities occurs in the first hour of aeration. Therefore the highest downwind concentrations may be expected during the first hour of aeration. While small facilities are unlikely to produce air concentrations exceeding the 0.210 ppm 24-hour TWA target level, there is a risk of exceedance at larger facilities. The ISCST3 model reproduced the general magnitude and pattern of the downwind concentrations measured during the commodity chamber fumigation study. These results indicate that the ISCST3 model is acceptable for use in characterizing the types of chamber fumigations expected to produce air concentrations exceeding the target level.

The large space fumigation study presented in this paper demonstrates that the release of methyl bromide from these fumigations (and related type of fumigations) occurs in two distinct phases: 1) treatment period leakage and 2) the aeration period release. It was shown that concentrations exceeding the 0.210 ppm 24-hour TWA target level can occur as a result of passive leakage during the treatment period from a sealed building. During the first hour of aeration, exhaust stack concentrations were very high and a large mass of methyl bromide was released during this period. The wind direction was constant and atmospheric conditions were neutral. As a result, some concentrations measured during the first hour of aeration exceed the 0.210 ppm 24-hour TWA target level. The ISCST3 model was shown to adequately reproduce the magnitude and pattern of downwind concentrations associated with both the treatment period and the aeration period. These results indicate that the ISCST3 model is acceptable for use in characterizing the types of fumigations expected to produce air concentrations exceeding the target level for both the treatment and aeration phases.

Acknowledgments

We would like to thank Paul Lee of the California Department of Food and Agriculture Chemistry Laboratory for performing all of the chemical analysis in these studies.

References

(1) California Department of Pesticide Regulation (CDPR). 1995. Annual pesticide use report. Sacramento, CA 95814-5624.
(2) Nelson , L. 1992. Memorandum to Jim Wells, dated February 11, 1992. Methyl Bromide preliminary risk characterization. Department of Pesticide Regulation, Sacramento CA 95814-5624.
(3) Andrews, C. 1994. Letter to County Agricultural Commissioners dated September 14, 1994. "Suggested permit conditions for Methyl Bromide commodity permit conditions." Department of Pesticide Regulation, Sacramento, CA 95814-5624.
(4) Okamura, D. 1992. Letter to County Agricultural Commissioners dated December 15, 1992. "Final permit conditions for Methyl Bromide soil injection fumigation." Department of Pesticide Regulation, Sacramento, CA 95814-5624.
(5) U.S. Environmental Protection Agency. 1993 Guideline on Air Quality Models (Revised) (1986), Appendix W of 40 CFR Part 51 with Supplement A (7/87) and Supplement B (2/93). EPA-450/2-78-027R.
(6) U.S. Environmental Protection Agency. 1995. User's guide for the Industrial Source Complex (ISC3) dispersion models, Volumes I&II. EPA-450/B-95-003a,b.
(7) Tikvart, J.A., 1993. Memorandum to Ken Eng, Chief, Air Compliance Branch, Region II, dated July 9, 1993. Proposal for calculating plume rise for stacks with horizontal releases or rain caps for Cookson Pigment, Newark, New Jersey. U.S. EPA SCRAM Bulletin Board, CFYM#89.TXT.

Chapter 15

Determination of Methyl Bromide in Air Resulting from Pest Control Fumigations

James E. Woodrow, Puttanna S. Honaganahalli, and James N. Seiber

Center for Environmental Sciences and Engineering and Department of Environmental and Resource Sciences, Mail Stop 199, University of Nevada, Reno, NV 89557–0187

A method for measuring residues of methyl bromide in air entails concentrating the fumigant on charcoal from an air-stream at a flow rate of ≤100 mL/min, desorbing the trapped material with benzyl alcohol solvent in a sealed vial at 60-110°C for 10-15 min, and then sampling the equilibrated vapor for gas chromatographic assay using electron-capture detection. The desorbed vapor is chromatographed on a 27 m x 0.32 mm (id) porous-layer open tubular column, on which methyl bromide has a retention time of about 6 min at 90°C and at a carrier gas flow rate of about 3.5 mL/min. Using this method, standard curves were linear over at least three orders of magnitude and a practical limit of detection for field air was less than 20 ng/m^3 (<5 ppt). Because of the possibility of methyl bromide hydrolysis on charcoal and uneven (or inconsistent) data precision using charcoal air sampling, we discuss the use of a polymeric adsorbent for air sampling and compare it with charcoal in terms of methyl bromide trapping, stability, and data precision.

Methyl bromide is extensively used in agriculture as a fumigant to control nematodes, weeds, and fungi in soil (e.g., 300-400 kg/ha is typically used in strawberry culture in California) and insect pests in harvested grains and nuts. Given its low boiling point (3.8°C) and high vapor pressure (~217 kPa at 25°C), methyl bromide will readily diffuse if not rigorously contained. When used as a soil fumigant, where the material is injected into the soil and immediately covered with a plastic tarp, significant amounts will escape (1-4); subsequent tarp removal will result in further releases to the atmosphere. Venting of fumigation sheds will also result in short-term releases of relatively high concentrations (>80 ppm) to the atmosphere (5). Of concern are exposures for field workers who apply the chemical, for those who work in and around previously treated fields, and for those who reside in regions near treated fields where chronic exposures to slowly released vapors may occur (6). The time-weighted average (8 hr/day, 40 hr/week) threshold limit value (TLV) for methyl bromide in air for exposed workers has been set at 1 ppm (~5 mg/m^3) (7); adjusted to a 24 hr day, 7

0097–6156/96/0652–0189$15.00/0
© 1996 American Chemical Society

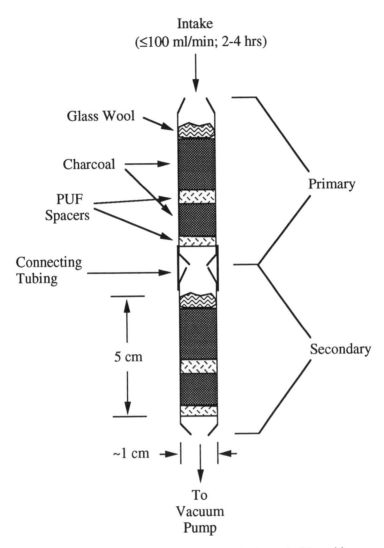

Figure 1. Charcoal tube air sampling train for methyl bromide.

day week, the TLV for residents in a region of use would be less than 0.25 ppm (~1 mg/m^3). These levels are recommended to prevent serious health effects. In general, halocarbons at ppm levels in air will affect the central nervous system and cause liver and kidney dysfunction (*8-11*). Of further, and perhaps greater, concern is the ozone depletion potential of methyl bromide if its vapors survive atmospheric dissipation processes long enough to enter the stratosphere. There is sufficient concern in this regard that methyl bromide has recently been categorized as a Class 1 ozone-depleting chemical by the U.S. Environmental Protection Agency and has been slated for at least a 25% reduction in use or outright ban by the Year 2000 (*12*). Thus, it becomes imperative that a simple and fast, yet accurate, method be available to determine methyl bromide in air to satisfy the following goals: 1) To provide input data for measuring evaporation rates from treated fields; 2) to provide data for exposure and risk assessment; 3) for comparison with dispersion models in discerning routes of dissipation; and 4) to contribute to the assessment of ozone depletion potential.

Methods for determining methyl bromide in air vary widely. Published methods include in-situ measurements of methyl bromide (*13*), and the use of steel canisters (*14-17*), septum-sealed vials (*1,2*), and solid adsorbents (e.g., Tenax GC, charcoal) (*3,4,18-20*) to trap methyl bromide for subsequent analysis in the laboratory. Studies concerned with the ambient levels of methyl bromide in the upper atmosphere and with atmosphere-ocean gaseous flux typically use evacuated steel canisters. For agricultural and structural fumigation uses, adsorbent trapping and in-situ spectrophotometric methods have been used. Adsorbent trapping techniques lend themselves to cumulative, time-averaged sampling common to many dissipation and environmental fate studies.

Post-sampling analytical techniques include the following: 1) Gas chromatography (GC) or gas chromatography/mass spectrometry (GC/MS) after cryofocusing samples from canisters and thermal desorption of adsorbents (e.g., Tenax GC) (*15,17-19*); 2) direct injection into GC or GC/MS from canisters, sealed vials and containers (*1,2,21,22*), and fumigation chambers using gas-tight syringes (*22*); and 3) headspace GC of equilibrated vapor for residues desorbed from adsorbents using solvents (e.g., benzyl alcohol with charcoal) (*3,4*), or direct injection into GC of solvent-desorbed residues (e.g., ethyl acetate with charcoal) (*20*).

We describe here a sampling and analytical approach that makes use of charcoal sampling of methyl bromide vapors followed by benzyl alcohol desorption in a sealed vial for vapor analysis. This approach was designed to handle large numbers of samples through automating some critical steps of the analysis, and the result is a method that allows around-the-clock operation with a minimum of operator attention. We compare charcoal adsorption with polymeric adsorbents in terms of trapping and stability of methyl bromide and the precision of the analytical data.

Sampling and Analysis Methods

To trap methyl bromide from air, we commonly use a sampling train consisting of two charcoal-filled glass tubes connected in series (Figure 1); the sampling tubes contain one gram each of coconut-based charcoal (Lot 120; SKC-West, Fullerton, CA). Sampling flow rates are typically ≤100 mL/min and sampling duration is usually 2-4 hours. The back, secondary tube is used to trap any residues that might break through the front, primary tube; if methyl bromide residues in the secondary tube equal or exceed 25% of the total residues (primary plus secondary), then the concentration in air should be reported as a "greater than" number. Immediately after sampling, each tube is capped and placed on dry ice for transport to the laboratory freezer, where the samples are stored at -20°C to await analysis.

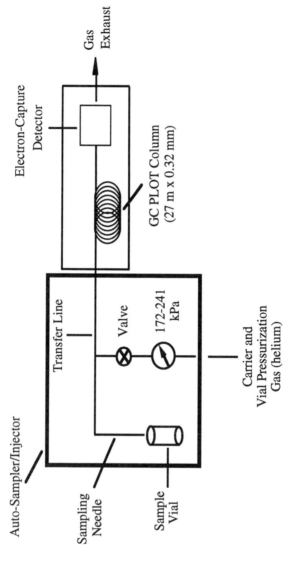

Figure 2. Schematic of the headspace auto-sampler/injector and gas chromatograph for methyl bromide analysis.

For methyl bromide determination, each tube is emptied into a 22 mL glass headspace vial (Perkin-Elmer, Norwalk, CT), 2.5 mL benzyl alcohol (Aldrich Chemical Co., Milwaukee, WI) is added, and the vial is sealed with a crimped cap containing a Teflon-lined butyl rubber septum (Perkin-Elmer). Each sealed vial is thermostated at 60-110°C for 10-15 min, pressurized to 172-241 kPa (gauge) for 0.5 min, and the equilibrated vapor sampled for 0.01-0.2 min. Ranges of the various conditions are used depending on the amount of methyl bromide in the samples and the desired sensitivity. The sampled vapor aliquot is chromatographed on a 27 m x 0.32 mm (id) PoraPlot Q porous layer open tubular (PLOT) column (ChromPak, The Netherlands) and methyl bromide detected using a ^{63}Ni electron-capture detector (ECD). The PLOT column and ECD are contained in a Perkin-Elmer Autosystem gas chromatograph connected to a Perkin-Elmer HS-40 autosampler, where the samples are processed (Figure 2). The sampling needle and transfer line (deactivated, uncoated fused silica capillary [0.32 mm id]) are both maintained at 170°C and the ECD is set at 350°C. Using this system, typical carrier gas (helium) flow rates through the PLOT column were about 3.5 mL/min, giving a methyl bromide retention time on this column of about 6 min at 90°C (Figure 3). With this column flow rate and a sampling time of 0.01 min, 35 μL of vapor would be injected onto the column (i.e., 0.01 min x 3.5 mL/min); with an injection time of 0.2 min, 700 μL vapor would be injected. About 10 min after injection, the temperature is programmed in steps to 210°C to clear the column for the next injection. Analysis cycle time (injection-to-injection) was about 26 min, suggesting that about 55 samples could be processed in a 24 hour period.

As an alternative to charcoal sampling, we selected Porapak Q (Supelco, Bellefonte, PA), an ethylvinylbenzene-divinylbenzene copolymer, for comparison. We thermally desorbed methyl bromide residues from this material at 110°C using the headspace instrument described above to generate standard curves and to determine trapping efficiency for 3 mL (1.02 g) aliquots of the adsorbent used to sample 4.7 pg/mL and 1,872 pg/mL standard atmospheres contained in 100 L Tedlar bags. In separate tests, flow rates were set at 10 mL/min and 20.8 mL/min for 60 min. Charcoal sampling and analysis, described above, were also used to determine methyl bromide in the standard atmospheres at these same flow rates for direct comparison with the Porapak Q material.

Results and Discussion

Charcoal adsorbent. We originally selected charcoal, rather than a polymeric adsorbent, as the trapping medium for methyl bromide in air because of the relatively high breakthrough volumes for charcoal. For example, Krost et al. (19) determined that for BPL grade charcoal, breakthrough volumes (L/g) ranged from 98 at 10°C to 25 at ~38°C. Use of charcoal at ≤100 mL/min flow for four hours (24 L) placed our method well within this range. This is in contrast to polymeric adsorbents, such as Tenax-GC, that have breakthrough volumes typically of the order of about 1 L/g, evaluated under similar conditions (18,19). Charcoal is a superior adsorbent because it contains impurities (i.e., metal salts, etc.) that act as adsorptive sites to chemisorb methyl bromide.   Polymeric adsorbents (e.g., Tenax-GC, XAD-2, XAD-4, polyurethane foam), commonly used to trap organics of low to intermediate volatility from air, rely mainly on van der Waals dispersive (electrostatic) forces for trapping. Since trapping of vapors by adsorbents is proportional to the relative vapor density, the surface potential of polymeric adsorbents is not great enough to trap methyl bromide as well as charcoal does because of the high vapor pressure of methyl bromide. For charcoal, trapping efficiency (amount trapped relative to the amount in air) falls in the range 82-85% for a flow rate of 100 mL/min over a four-hour sampling period. Under

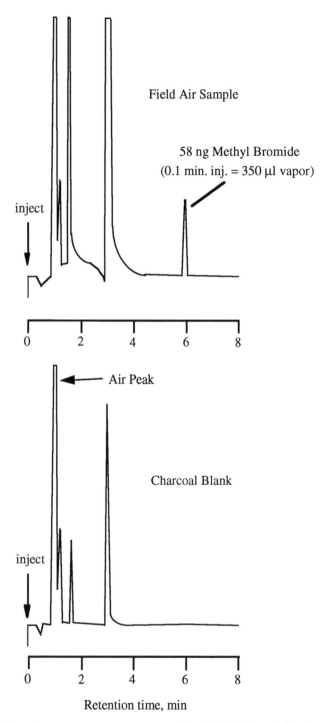

Figure 3. Gas chromatograms of methyl bromide in field air and of a charcoal blank.

the same conditions, the polymeric adsorbent XAD-4 (divinylbenzene copolymer; 20-50 mesh), for example, has a trapping efficiency of only a few percent. Charcoal is very effective in trapping methyl bromide from air and then releasing it with benzyl alcohol desorption. About 80-100% of the adsorbed methyl bromide is released from charcoal by benzyl alcohol, with at least 10% of the desorbed material residing in the equilibrated vapor at 60°C. Even so, the detection limit for methyl bromide is 0.5 ng on charcoal which is equivalent to about 20 ng/m^3 in air (100 mL/min flow for 4 hours), and even less for a sample temperature of 110°C. The gas chromatographic conditions gave a several minute window for methyl bromide elution (Figure 3). This window was virtually free of interferences, even at the 110°C maximum sample thermostating temperature; if methyl bromide residues were below the detection limit, the gas chromatogram was essentially a flat baseline. This clean window allowed us to generate standard curves that were linear over several orders of magnitude (Figure 4).

A draw-back to the use of charcoal is the fact that in the presence of water alkaline (pH > 10) conditions will result promoting methyl bromide hydrolysis (*23*):

$$CH_3Br + OH^- \text{--------} > CH_3OH + Br^-$$

At 25, 50, 60, and 70°C, the hydrolysis half-lives ($t_{1/2}$) were 11 days and 13.6, 4.0, and 1.7 hours, respectively. At the -20°C freezer storage temperature, $t_{1/2}$ was greater than a year (*23*). Therefore, for the three to five month storage period, 80-90% of the original methyl bromide was still intact. These results imply, however, that under high humidity and warm conditions (e.g., fog, rain, marine air) it may be necessary to pre-treat the air to remove the moisture prior to charcoal adsorption to avoid hydrolysis during sampling, or to employ a non-alkaline, hydrophobic (i.e., polymeric) adsorbent or use other techniques entirely (e.g., canisters, in-situ measurements). Other investigators have suggested correcting for methyl bromide loss during sampling with charcoal by taking the relative humidity and temperature into account (*23*).

The precision of the headspace analytical instrument was determined by measuring variation in response of the air peak (Figure 3). In doing so for 46 determinations, the range of absolute variation (difference between each determination and the mean of the set of determinations) was 0.03-8.9%, with a mean value of 1.9%. The manufacturer's claim is about 1% variation. By contrast, the absolute variation in the methyl bromide response for 44 samples of standard-spiked charcoal fell in the range 0.44-43%, with a mean value of 11.6% (26 out of the 44 determinations were less than 10%). Similarly, the absolute variation for 46 (23 pairs) collocated methyl bromide samples trapped on charcoal under field conditions (Table 1) was in the range 0.3-53%, with a mean value of 11.6% (25 out of the 46 determinations were less than 10%). In all cases for methyl bromide on charcoal, variation appeared to be random and was not a function of residue level. These results suggest that the charcoal itself introduces significant variation in methyl bromide determination. This may be due to a lack of uniformity in the properties of the charcoal, tube-to-tube, leading to a variation in the ability of the charcoal to trap methyl bromide and subsequently release it for analysis.

The observed variability in sampling methyl bromide with charcoal and the latter's tendency to promote methyl bromide hydrolysis in moist environments suggest that it might be prudent to examine other sampling methods as possible refinements to or replacements for charcoal air sampling. With regard to the moisture problem, a possibility would be to add a calcium chloride or phosphorus pentoxide moisture trap upstream of the charcoal sampling train. This approach has been used to prepare marine air samples in canisters for analysis (*17*). The outright replacement of charcoal

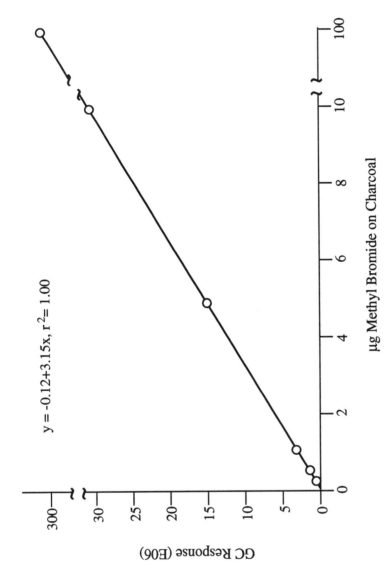

Figure 4. Standard curve for methyl bromide spiked to charcoal.

Table 1. Methyl bromide concentration in field air for collocated samplers

Sampling Station[b]	Collocated pairs[a] I	II	Sampling Station[b]	Collocated pairs[a] I	II
C	47.7	65.1	G	6.69	5.41
	19.5	15.3		2.80	2.63
	11.5	11.4		4.21	3.97
	1.13	1.71		0.422	0.440
D	7.86	13.4	H	2.06	2.84
	35.6	47.7		1.35	0.873
	9.48	9.54		7.35	6.03
	11.3	10.2		4.04	4.16
E	4.36	10.5	I	0.359	0.329
	7.48	9.65		5.29	4.40
	3.99	5.66		2.49	2.34
	18.4	19.4		0.800	0.736
F	3.47	4.07	J	4.21	5.07
	6.91	6.27		0.480	0.991
	3.21	4.69		2.13	1.72
	0.168	0.544		2.39	1.75
	2.59	3.53	K	1.17	1.03
	23.9	25.0		3.15	3.55
	1.87	1.27		0.448	0.660
	10.9	12.2		22.1	18.7
	2.93	2.82	L	2.65	3.38
	5.81	5.19		18.7	12.0
	0.476	0.473		2.33	2.41

[a] Entries are concentrations in air ($\mu g/m^3$).

[b] Figure 1; Chapter: Seiber et al., Flux, dispersion characteristics, and sinks for airborne methyl bromide downwind of a treated agricultural field.

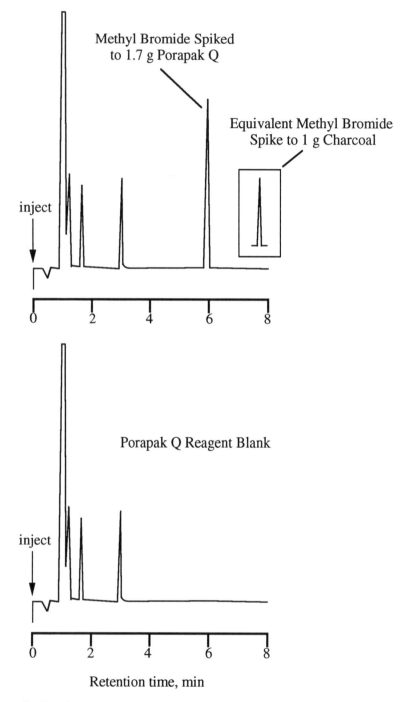

Figure 5. Gas chromatograms of methyl bromide spiked to Porapak Q and of a reagent blank.

with polymeric adsorbents is a possibility, except for the much lower breakthrough volumes associated with these adsorbents (*18,19*). However, the increased methyl bromide stability that would be expected with polymeric adsorbents and the improved data precision, discussed below, may be enough to off-set the relatively poor breakthrough volumes.

Polymeric adsorbents. A common polymeric adsorbent used for trapping volatile organics, such as methyl bromide, is Tenax-GC, a polymer of 2,6-diphenyl-*p*-phenylene oxide. This material has a very low affinity for water and has high thermal stability, which makes it useful for thermal desorption applications. Its breakthrough volume for methyl bromide has been estimated to be about 1 L/g (*18,19*) for 0.13 g of the adsorbent (40-50 mesh) at a flow rate range of 5-600 mL/min, temperature of 20°C, and a methyl bromide vapor density of less than 250 mg/m^3 (*18*). The same break-through volume for methyl bromide was determined for approximately 2 g of Tenax-GC (60-80 mesh) at a flow rate range of 5-9 L/min and temperature of 21°C; for 10°C and 32°C, the breakthrough volumes were >1.3 L/g and <0.5 L/g, respectively (*19*). These results show that breakthrough volume is related to such factors as sampling flow rate and temperature; the vapor density of methyl bromide and mass of adsorbent are also factors that affect breakthrough volume (*24*). For quantitative sampling using polymeric adsorbents, it is important to find the right combination of these factors to allow for some flexibility, especially with regard to ranges of vapor concentrations and temperatures.

We recently examined Porapak Q, the same material used in the GC PLOT column for methyl bromide analysis, as another possible replacement for charcoal air sampling. Porapak Q appears to be hydrophobic and it does not change the pH of added water; however, long-term stability of methyl bromide on this material at ambient and freezer temperatures would have to be assessed. Furthermore, thermal desorption of methyl bromide residues on Porapak Q at 110°C gives an almost three-fold greater GC response than solvent-desorbed residues on charcoal at the same spiking level and temperature (Figure 5). Finally, for tests run side-by-side, the absolute percent variation for methyl bromide residues on charcoal solvent-desorbed at 110°C fell in the range 1-18.4% (ave = 12.2%, n = 3); this compares with the much larger data set discussed above. For Porapak Q, spiked at the same level and thermally desorbed at 110°C, the range was 0.5-4.5% (ave = 3.0%, n = 3).

Preliminary air sampling results obtained from a series of tests in our laboratory at 25°C indicated that about 3 mL (1.02 g) of 100/120 mesh Porapak Q trapped 100% of the methyl bromide from 4.7 pg/mL (µg/m^3) and 1,872 pg/mL standard vapors at about 10 mL/min flow for one hour of sampling. When the flow rate was approximately doubled to 20.8 mL/min for the 1,872 pg/mL standard vapor, however, the recovery fell to about 48%. These results, for vapor concentrations that span the range typically found near treated fields (≤1 km) for a period of up to about a week after application, suggest that quantitative sampling could be achieved in the field by using two tubes in series each containing 6 mL (2.04 g) of 100/120 mesh Porapak Q at a flow rate of 20 mL/min for about a two-hour sampling period (2.4 L). This approach is supported by preliminary results with a single tube containing 5 mL (1.7 g) Porapak Q which quantitatively trapped methyl bromide from a 4.7 pg/mL vapor at 20 mL/min for about one hour. Greater volumes of air could be sampled over a two-hour period by proportionately increasing together the flow rate and amount of Porapak Q (*24*). However, a limiting factor in this case is pressure drop through the sampling train; too great a pressure drop could result in vacuum stripping of trapped residues. It may be prudent in this case to use coarser material (e.g., 60/80 or 80/100 mesh) to help minimize pressure drop. Conversely, too low a flow rate, to minimize pressure drop, could lead to poor theoretical plates for the adsorbent and, thus, poor trapping (*18*).

Conclusions

Coconut-based charcoal efficiently trapped methyl bromide from air and quantitatively released the residues using benzyl alcohol desorption. The desorbed residues were readily quantitated by sampling the equilibrated vapor in sealed vials using automated headspace gas chromatography. Sampling at a flow rate of 100 mL/min for 4 hours and thermostating the desorbed samples at 110°C, we were able to achieve a limit of detection (LOD) of less than 20 ng/m^3 (5 ppt). While average variation in instrument response was about 2%, variation in methyl bromide response, for both standard-spiked and field samples, averaged 11.6%, with ranges of 0.44-43% and 0.3-53%, respectively. It is surmised that charcoal introduces significant variability to methyl bromide analysis; this variability may reflect variability in trapping, desorption, and stability. With regard to the latter, moist charcoal will promote the hydrolysis of methyl bromide residues, and because of this there is some concern regarding the sampling of warm, moist atmospheres. However for long-term storage, hydrolysis can be minimized by maintaining samples at -20°C prior to analysis (>90% of initial residues were recovered after 3-5 months).

Compared to charcoal, polymeric adsorbents, such as Tenax-GC and Porapak Q, have advantages in terms of greater methyl bromide stability and better data precision. However, these polymeric adsorbents typically have poorer breakthrough volumes, requiring lower flow rates and shorter sampling periods for quantitative trapping. On the other hand, methyl bromide residues can be thermally desorbed from Porapak Q, and probably from Tenax-GC as well, for headspace analysis giving the polymeric adsorbent about a 3-to-1 advantage over solvent-desorbed charcoal under the same conditions. For example, air sampling at 30 mL/min for two hours using a sampling train of two tubes in series, each containing 9 mL (3.06 g) Porapak Q, should be about equivalent to a two-tube charcoal sampling train (1 g charcoal in each tube) operated at 100 mL/min for the same period of time. This takes into account the approximate 3-to-1 greater GC response of Porapak Q over charcoal. Furthermore, it should be possible to better charcoal's advantage in terms of LOD (i.e., ~20 ng/m^3 [5 ppt] for charcoal compared to ~47 ng/m^3 [12 ppt] for Porapak Q) by thermally desorbing Porapak Q, as before, but using an inert gas to sweep the entire desorbed residue into a concentrator (i.e., a cryofocusing unit) prior to gas chromatographic analysis. In this way, it should be possible to improve the LOD for Porapak Q to an order of magnitude better than that for charcoal (i.e., ~2 ng/m^3 [0.5 ppt]).

Literature Cited

1. Yagi, K.; Williams, J.; Wang, N.Y.; Cicerone, R.J. *Proc. Natl. Acad. Sci. USA*, **1993**, *90*, 8420-8423.
2. Yagi, K.; Williams, J.; Wang, N.Y.; Cicerone, R.J. *Science*, **1995**, *267*, 1979-1981.
3. Majewski, M.S.; McChesney, M.M.; Woodrow, J.E.; Pruger, J.H.; Seiber, J.N. *J. Environ. Qual.*, **1995**, *24*, 742-752.
4. Woodrow, J.E.; McChesney, M.M.; Seiber, J.N. *Anal. Chem.*, **1988**, *60*, 509-512.
5. Maddy, K.T.; Lowe, J.; Fredrickson, A.S. *Inhalation Exposure of Commodity Handlers to Methyl Bromide in Yolo County, California, October, 1993*. Report HS-1168; California Department of Food and Agriculture: Sacramento, CA, June 4, 1984.

6. Guillemin, M.P.; Hillier, R.S.; Bernhard, C.A. *Ann. Occup. Hyg.*, **1990**, *34*, 591-607.
7. ACGIH. *Documentation of the Threshold Limit Values and Biological Exposure Indices*; American Conference of Governmental Industrial Hygienists: Cincinnati, OH, 1991.
8. Herzstein, J.; Cullen, M.R. *Am. J. Ind. Med.*, **1990**, *17*, 321-326.
9. Bishop, C.M. *Occup. Med.*, **1992**, *42*, 107-109.
10. Fuortes, L.J. *Vet. Hum. Toxicol.*, **1992**, *34*, 240-246.
11. Hustinx, W.N.M.; van de Laar, R.T.H.; van Huffelen, A.C.; Verwey, J.C.; Meulenbelt, J.; Savelkoul, T.J.F. *Br. J. Ind. Med.*, **1993**, *50*, 155-159.
12. UNEP. United Nations Environment Program. Montreal Protocol Synthesis Report of the Methyl Bromide Interim Scientific Assessment and Methyl Bromide Interim Technology and Economic Assessment, 1992.
13. Green, M.; Seiber, J.N.; Biermann, H.W. *Proc. Spie. Int. Soc. Opt. Eng.*, **1993**, *1716*, 157-164.
14. Rasmussen, R.A.; Khalil, M.A.K.; Fox, R.J. *Geophys. Res Lett.*, **1983**, *10*, 144-147.
15. Khalil, M.A.K.; Rasmussen, R.A. *Antarct. J. U.S.*, **1992**, *27*, 267-269.
16. Sheridan, P.J.; Schnell, R.C.; Zoller, W.H.; Carlson, N.D.; Rasmussen, R.A.; Harris, J.M.; Sievering, H. *Atmos. Environ., Part A*, **1993**, *27A*, 2839-2849.
17. Lobert, J.M.; Butler, J.H.; Montzka, S.A.; Geller, L.S.; Myers, R.C.; Elkins, J.W. *Science*, **1995**, *267*, 1002-1005.
18. Brown, R.H.; Purnell, C.J. *J. Chromatogr.*, **1979**, *178*, 79-90.
19. Krost, K.J.; Pellizzari, E.D.; Walburn, S.G.; Hubbard, S.A. *Anal. Chem.*, **1982**, *54*, 810-817.
20. National Institute of Occupational Safety and Health. *NIOSH Manual of Analytical Methods*; Eller, P.M., Ed.; 3d ed.; NIOSH: Cincinnati, OH, 1984.
21. King, J.R.; Benschoter, C.A.; Burditt, A.K. *J. Agric. Food Chem.*, **1981**, *29*, 1003-1005.
22. Dumas, T.; Bond, E.J. *J. Agric. Food Chem.*, **1985**, *33*, 276-278.
23. Gan, J.; Anderson, M.A.; Yates, M.V.; Spencer, W.F.; Yates, S.R. *J. Agric. Food Chem.*, **1995**, *43*, 1361-1367.
24. Ness, S.A. *Air Monitoring for Toxic Exposures: An Integrated Approach*; Van Nostrand Reinhold: New York, NY, 1991; pp 51-68.

Chapter 16

Time-Resolved Air Monitoring Using Fourier Transform Infrared Spectroscopy

Heinz W. Biermann

Environmental Monitoring and Pest Management, Department
of Pesticide Regulation, California Environmental Protection Agency,
1020 N Street, Sacramento, CA 95814–5624

Two major advantages of Fourier transform infrared (FTIR) spectroscopy
are the capabilities to perform air analyses *in situ* and to obtain data at high
time resolutions. Taking air measurements *in situ* allows this technique to
bypass most error sources associated with conventional sampling. This
paper concentrates mainly on the description of the instrumentation and the
data analysis procedures used. Three data sets obtained with this FTIR
system previously are mentioned as examples for the information that can
be gained with this technique. In two cases, a 100 m folded optical path was
used to measure methyl bromide concentrations after structural fumigations
of residential homes and after commodity fumigation in warehouses. The
time resolution was 15 min with a detection limit of about 0.2 ppm. In
addition, trying to assess the capability of this FTIR spectrometer to
determine flux, water vapor concentrations were measured with a
four-meter folded path length at a time resolution of 0.6 seconds.

Fourier transform infrared (FTIR) spectroscopy has been used for about 20 years to
monitor air pollutants (1-5). The technique is not suited to measure the majority of
pesticides because of their low vapor pressures and consequently low air concentrations.
Fumigants, on the other hand, are excellent target compounds for FTIR measurements
because of their relatively high vapor pressures.

The basic principle of this technique is to send an infrared light beam through the air and
then to monitor the intensity of the transmitted light. Any chemical in the light path will
lower the intensity at characteristic wavelengths due to absorption. Gas mixtures (like
methyl bromide/chloropicrin, for example) can be analyzed as easily as single components.
Concentrations can be derived from the light intensity measurements through the Beer-
Lambert law: the logarithm of the incident versus resultant light intensity is equal to the
product of the absorption path length, the concentration of the chemical and a calibration
factor.

0097–6156/96/0652–0202$15.00/0
© 1996 American Chemical Society

Compared to standard air sampling methods that pull air through a cartridge containing an adsorbent and then analyze the adsorbed material by GC methods, the FTIR technique holds a major advantage. Air concentrations can be determined *in situ* and at high time resolution. Thus, this method bypasses a number of problems related to sampling and chemical analysis such as sampling efficiency, breakthrough and extraction efficiency. A major disadvantage of the FTIR is the relatively high detection limit. Because it is an *in situ* technique, it cannot accumulate the target compounds over an extended time period as trapping methods do.

Another important difference is that *in situ* absorption spectroscopy can detect gaseous material only. Chemicals adsorbed onto particulate matter will not be detected. Large amounts of particulate matter in the path of the light beam, however, reduce the intensity of the transmitted light due to scatter.

Instrumentation

The instrument is based on a commercial FTIR spectrometer (KVB Analect RFX-75) with a KBr beam splitter and a liquid N_2-cooled broad band HgTeCd detector interfaced to external multiple reflection optical systems based on the design by White (6). Figure 1 shows the layout of the instrument and the external optics using a commercial 1 m base path system (Infrared Analysis, Inc.). For ambient air measurements, the mirror assembly is used without any enclosure. The mirrors can be inserted into an evacuable glass jacket to obtain clean reference spectra. The other two multiple reflection systems currently in operation use custom optics with base distances of 2.5 m and 10 m, respectively. With these distances between the mirrors, total absorption path lengths of 26 m, 100 m and 260 m can be achieved.

These total path lengths do not represent the maximum possible, but rather correspond to a setting that can be maintained over extended time periods under field conditions. Even under optimal atmospheric conditions, however, the diurnal temperature changes cause slight movements in the optical alignment that have to be corrected frequently. To correct for changes in the alignment, an external HeNe laser is used to check the aim of the mirrors periodically. However, extreme adverse atmospheric conditions, like dense dust or fog, as well as heavy rain, make it impossible to transmit the light beam through the atmosphere.

Data analysis

The raw data acquired by the instrument are in the form of an interferogram and have to be converted to absorption spectra using the Fourier transform algorithm. Figure 2 shows two sample spectra after the transformation: a methyl bromide reference spectrum taken in a cuvette with a path length of 15 cm and an ambient air background taken with an absorption path length of 100 m. At the long path lengths employed, water and carbon dioxide vapors present in the atmosphere block out large parts of the spectrum. One has to find spectral regions that have no or minimal interference from these ubiquitous compounds. In the case of methyl bromide, only the wavelength region around 950 cm^{-1} is useable for quantitation. Concentrations can be calculated from these spectra using the Lambert-Beer law:

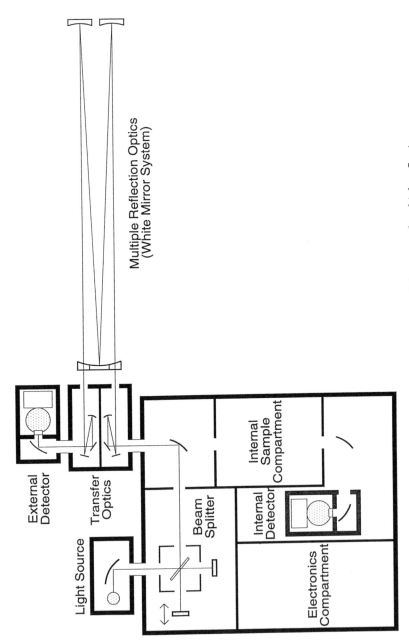

Figure 1: Schematics of the FTIR system and the external multiple reflection optics.

$$\log(\frac{I_0}{I})=a*c*l$$

where I_0 and I are the incident and resultant light intensities, a is the absorption coefficient, c the concentration and l the length of the light path. Thus, the concentrations calculated from absorption measurements are averages over the length of the light beam.

While it is easy to generate a clean background spectrum when one uses enclosed sampling volumes, because enclosed volumes can be evacuated or flushed with clean air, it is impossible to get an *in situ* reference for the light intensity without any absorption from ambient air. The solution is to measure the difference between a peak minimum and maximum instead of the absolute intensities:

$$\log(\frac{I'_0}{I})=a'*c*l$$

where I'_0 is the light intensity at the base of an absorption line, I the intensity at the minimum of that absorption line, a' is the differential absorption coefficient, c the concentration and l the length of the light path. Because concentrations are proportional to the logarithm of the light intensity ratio, it is convenient to show spectra in a linearized form using absorbance, log (I_0/I), on the y-axis. In infrared spectroscopy it is also a convention to plot wavenumbers (1/cm) on the x-axis instead of a wavelength scale. Because wavenumbers are inversely proportional to the wavelength, the wavenumber values are generally plotted with the highest value on the left end of the graph.

There are numerous peaks in a reference spectrum that could be used to quantify the amount of methyl bromide in air samples; but it would be a big loss in specificity if one were to use just a single peak in the analysis procedure. In order to be sure that the observed absorption is really due to methyl bromide, all peaks in a sample spectrum that are not obscured by water or carbon dioxide have to match with their relative intensities to this reference. This matching can be done numerically by fitting the reference spectrum to an air spectrum using a least-squares algorithm.

A complication arises from the presence of additional absorption lines from other chemicals in the air (notably water). As the number and intensities of unrelated absorption lines in a spectrum increases, this fitting procedure would yield increasingly erroneous results. If the number of additional lines is small, one solution is to exclude those areas from the least-squares fit. But if there is a large number of extraneous lines, reasonable results can be obtained only if those lines are included in the fitting procedure. Thus the algorithm has to be modified to approximate any given air spectrum as a linear combination of reference and background spectra.

In Figure 3, for example, trace a) is an air spectrum obtained after a fumigation with methyl bromide. The following three reference spectra were used in the fitting procedure: b) an ambient background spectrum taken before the fumigation (all water lines have been removed in a previous fitting procedure), c) a water reference spectrum and e) a methyl bromide reference spectrum. Because the unknown sample will be approximated by a linear combination of reference spectra, each set of absorption lines due to a single chemical can appear in only one of the reference spectra. If the same information is present

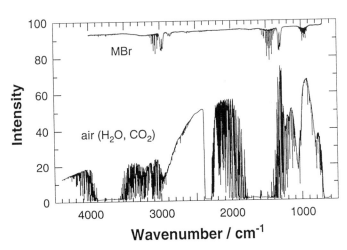

Figure 2: Sample FTIR spectra; top: methyl bromide reference (15 cm path length), bottom: ambient air (100 m path length).

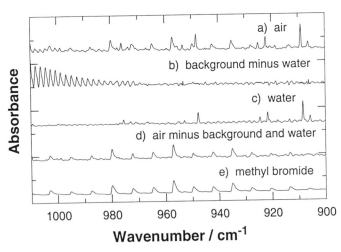

Figure 3: Example of data analysis procedure; trace a): ambient air sample after fumigation with methyl bromide, trace b): background air before fumigation, minus water, trace c): water reference spectrum, trace d): air sample minus background and water, trace e): methyl bromide reference spectrum.

in more than one reference spectrum (like water lines being present in two of the spectra), the algorithm cannot calculate a unique value for the intensity of these lines, and the fitting results become meaningless. Therefore, the water lines originally present in the ambient background shown in Figure 3 were removed from that spectrum first.

To demonstrate the procedure more clearly, Figure 3 also includes an intermediate step not taken during routine analysis. The air spectrum shown in the top trace is fitted first with the background and water spectra only. This stepwise approximation works well in this case because methyl bromide is only a minor component in the ambient air spectrum. Thus, trace d) shows the residual spectrum after subtracting the least-squares fit of the linear combination of traces b) and c) from trace a). Comparison with the methyl bromide reference shown in trace e) shows a very clear signature of methyl bromide.

From the scaling factor obtained by the least-squares procedure one can calculate the ambient concentration using the formula:

$$C_s = f_{ls} * C_r * \frac{l_r}{l_s}$$

where C_s, C_r and l_s, l_r are the concentrations and absorption path lengths in the sample and the reference, respectively, and f_{ls} is the scaling factor from the least-squares routine.

Contrary to conventional air analyses, where there are numerous error sources in the field sampling, transport/storage and quantitation steps, this technique has only four sources of error between measurement of the light intensity and the final concentration. For the two path lengths (l_s and l_r), the associated error can easily be held to less than 1%. And, barring any major systematic errors in the calibration, the uncertainty of the concentration in the reference spectrum will usually be below 5%. It is quite feasible to lower this uncertainty further because the reference spectra stay valid as long as the operating parameters of the instrument are not modified. For example, the methyl bromide reference used in these analyses was taken over two years before the ambient data. Yet the overlay of traces d) and e) from Figure 3, enlarged in Figure 4, shows that the processed ambient spectrum and the laboratory reference are virtually indistinguishable and would be identical if it were not for the excess noise in the ambient data.

Applications

Indoor methyl bromide concentrations have been determined using this technique after structural fumigations inside homes (7) and after commodity fumigations inside warehouses (8). Figures 5 and 6 show some of the results of these studies in the form of time-concentration profiles. In both cases, air spectra were acquired with a time resolution of 15 minutes. This resolution makes it possible to see short term fluctuations in the methyl bromide concentration.

Figure 5 shows a significant rise in methyl bromide after the instrument had been set up inside the house with the doors and windows closed, even though the house had been declared safe (<3 ppm by Drager tube). The concentration dropped sharply whenever the ventilation was increased by opening either the doors or doors and windows.

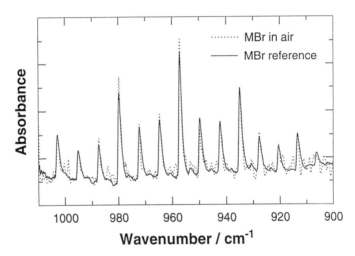

Figure 4: Overlay of residual lines after background removal from an ambient air spectrum (dotted line) and a scaled methyl bromide reference spectrum acquired over two years earlier (solid line).

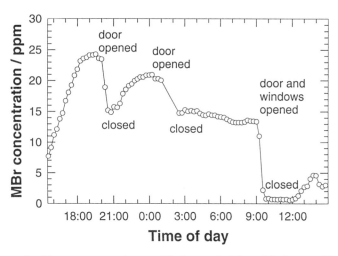

Figure 5: Time-concentration profile for methyl bromide in a residential home, starting about six hours after the house was declared safe for reentry. Rapid changes in the concentration occurred when the air circulation was changed as indicated in the graph.

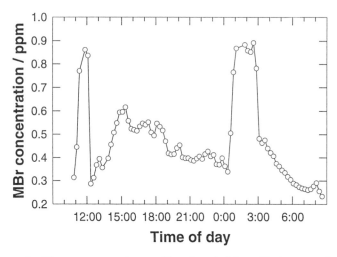

Figure 6: Time-concentration profile of methyl bromide in a warehouse, starting about 50 hours after the beginning of aeration. The two peaks in the profile correspond to times of reduced ventilation.

Similarly, Figure 6 shows unexpected structure in the methyl bromide concentration inside a warehouse about 50 hours after aeration. It turned out that the two peaks in the time-concentration profile were caused by reduced air circulation in the building during these times. These two examples clearly show the additional, useful information that can be gained from the high time resolution that this technique affords. As a result of these and other data, regulatory requirements have been revised. Buildings must now be aerated for an extended period and air concentrations measured before reoccupancy.

The 15 minute averaging interval during these indoor air studies was chosen just for convenience. The fastest data acquisition speed that this instrument is capable of is about 1.5 spectra per second. This makes it feasible to use this instrument for flux measurements. In collaboration with Prof. Kyaw Tha Paw U from the University of California, Davis, the FTIR was collocated with his fast response (10 Hz) humidity sensor (KH20 UVGA).

Figure 7 shows a comparison of the water vapor concentrations measured with the two instruments. Because of its faster response time, the UVGA instrument picks up higher frequency variations in the water vapor concentration. To make a comparison to the FTIR data more easily, the UVGA data were filtered digitally to simulate the time resolution of the slower FTIR. Considering that the UVGA gets its signal from a 10 cm path and the FTIR integrates over a 1 m path, the variations in the water concentration in the two signals are very similar. Even more remarkable, initial calculations based on a single 20 minute time period yielded flux values of $1.01 * 10^{-4}$ kg m^{-2} s^{-1} for the UVGA using the eddy covariance method and $1.00 * 10^{-4}$ kg m^{-2} s^{-1} for the FTIR using the surface renewal method (9).

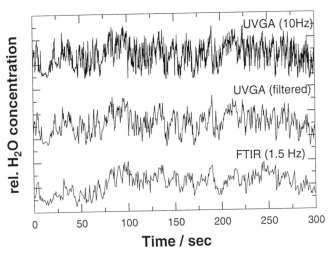

Figure 7: Comparison of water vapor concentrations between UVGA and FTIR. The center trace shows the UVGA data digitally filtered to the match the response time of the FTIR data.

Conclusions

Outdoor, *in situ* FTIR measurements are not a method of choice for most routine air monitoring. The equipment is quite expensive, and maintaining a reasonable signal strength over extended time periods can be quite challenging in the field. Despite these shortcomings, FTIR absorption spectroscopy can be a very useful and powerful tool to determine the influence of sources, sinks and mixing on short-term air concentrations. The technique yields concentration values quite directly, bypassing all the sources of uncertainty associated with the storage, handling and analysis of chemical samples. Another major advantage is that FTIR spectrometers can be operated at time resolutions unachievable by conventional sampling methods.

Acknowledgments

The author wishes to thank R. Segawa and B. Johnson of the Environmental Monitoring and Pest Management branch of CalEPA/DPR for coordinating and assisting in the studies mentioned here. I am also grateful to Prof. Kyaw Tha Paw U of the University of California, Davis for his collaboration and assistance in the acquisition and interpretation of the flux data.

References

1. Hanst, P. L.; Spectroscopic Methods for Air Pollution Measurement, *Adv. Environ. Sci. Technol.*,1971, *2*, 91.
2. Tuazon, E. C.; Graham, R. A.; Winer, A. M.; Easton, R. R.; Pitts, Jr., J. N.; Hanst, P. L.; A Kilometerpathlength Fourier-Transform Infrared System for the Study of Trace Pollutants in Ambient and Synthetic Atmospheres, *Atmos. Environ.*, 1978, *12*, 865.
3. Herget, W. F.; Bradsher, J. D.; Remote Measurement of Gaseous Pollutant Concentrations Using a Mobile Fourier Transform Interferometer System, *Applied Optics*, 1979, *18*, 3403.
4. Maker, P. D.; Niki, H.; Savage, C. M.; Breitenbach, L. P.; Fourier transform Infrared Analysis of Trace Gases in the Atmosphere, in *Monitoring Toxic Substances*, D. Schuetzle, Ed., ACS Symposium Series 94, American Chemical Society, Washington, DC, 1979.
5. Hanst, P. L.; Wong, N. W.; Bragin, J.; A Long-Path Infra-Red Study of Los Angeles Smog, *Atmos. Environ.*, 1982, *16*, 969.
6. White, J. U.; Very long optical paths in air, *J. Opt. Soc. Am.*, 1976, *66*, 411.
7. Green, M.; Biermann, H. W.; Seiber, J. N.; Long-path Fourier transform infrared spectroscopy for post-fumigation indoor air measurements, *Analusis*, 1992, *20*, 455.
8. Biermann, H. W.; Segawa, R.; Wofford, P.; Time-resolved measurements of post-fumigation methyl bromide concentrations in warehouses using long pathlength Fourier transform infrared spectroscopy, in preparation.
9. Paw U, K. T.; Biermann, H. W.; Johnson, B. R.; A comparison of FTIR and UVGA eddy covariance and surface renewal estimates of water vapor and carbon dioxide fluxes, preprints, 22nd Conference on Agricultural and Forest Meteorology, Atlanta, GA, January 1996, American Meteorological Society, Boston, MA, 1995.

Chapter 17

Determination of 1,3-Dichloropropene Degradates in Water and Soil by Capillary Gas Chromatography with Mass Spectrometric Detection

D. O. Duebelbeis, A. D. Thomas, S. E. Fisher, and G. E. Schelle

Global Environmental Chemistry Laboratory—Indianapolis Laboratory, DowElanco, 9330 Zionsville Road, Indianapolis, IN 46268–1054

Methods for the determination of *cis*- and *trans*- isomers of 3-chloroallyl alcohol (CAAL) and 3-chloroacrylic acid (CAAC) in water and soil were developed. Both CAAL and CAAC have been identified as major degradates of the soil fumigant, 1,3-dichloropropene (1,3-D). The methods required derivatization prior to quantitation by gas chromatography with mass spectrometric detection (GC/MS). CAAL residues were converted to the corresponding *cis*- and *trans*-3-chloroallyl isobutyl carbonates using isobutyl chloroformate. CAAC residues were converted to the corresponding *cis*- and *trans*-3-chloroacrylic acid *t*-butyldimethylsilyl esters using *N*-methyl-*N*-(*t*-butyldimethylsilyl)trifluoroacetamide (MTBSTFA). The GC/MS used electron impact ionization with selected ion monitoring (SIM) of two ions for each analyte. Recoveries for CAAL in surface water fortified at the 0.1 ng/mL level averaged 83% with a standard deviation (SD) of 7% for both the *cis*- and *trans*- isomers. Recoveries for CAAL in soil fortified at the 0.4 ng/g level averaged 88% with a SD of 10% for the *cis*- and 90% with a SD of 9% for the *trans*- isomers. Calculated limits of detection (LOD) for each CAAL isomer in surface water and soil were 0.02 ng/mL and 0.1 ng/g, respectively. Recoveries for CAAC in surface water fortified at the 0.05 ng/mL level averaged 89 and 90%, each with a SD of 6%, for the *cis*- and *trans*- isomers, respectively. Recoveries for CAAC in soil fortified at the 0.2 ng/g level averaged 83% with a SD of 11% for the *cis*- and 87% with a SD of 9% for the *trans*- isomers. Calculated LOD's for each CAAC isomer in surface water and soil were 0.009 ng/mL and 0.06 ng/g.

1,3-Dichloropropene (1,3-D) has seen wide use as a fumigant for the control of soil nematodes. In the presence of water, 1,3-D hydrolyzes to the 3-chloroallyl alcohol (CAAL) (*1*). Microbial action can convert the alcohol to the 3-chloroacrylic acid (CAAC) (*2-4*). 1,3-D is applied as a mixture of the *cis*- and *trans*- isomers and both isomers of each degradate have been observed (Figure 1).

0097–6156/96/0652–0212$15.00/0
© 1996 American Chemical Society

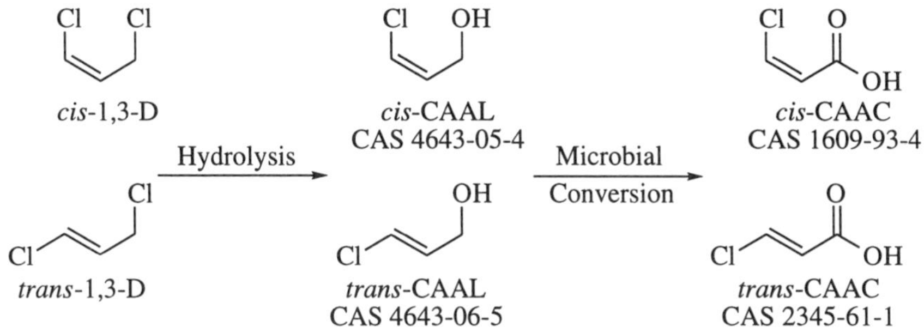

Cl   Cl
cis-1,3-D

Cl   OH
cis-CAAL
CAS 4643-05-4

Cl   O
               OH
cis-CAAC
CAS 1609-93-4

Hydrolysis →   Microbial Conversion →

Cl
Cl⁀   trans-1,3-D

OH
Cl⁀   trans-CAAL
CAS 4643-06-5

O
Cl⁀   OH
trans-CAAC
CAS 2345-61-1

Figure 1. Environmental degradation of 1,3-D in soil.

The following methods for the *cis-* and *trans-* isomers of CAAL and CAAC in water and soil were developed to support studies designed to assess the environmental fate of 1,3-D. Limits of quantitation (LOQ) of 0.05 to 0.10 ng/mL in water and 0.2 to 0.4 ng/g in soil were targeted. The procedures designed for sampling limited the aliquot available for water analysis to 40-mL. A 10-g sample size was utilized for soil analysis.

Rapid and sensitive determination of 1,3-D is achieved through purge and trap methodology; unfortunately, the increased water solubility of the polar degradates preclude this approach. A number of gas chromatographic (GC) methods have been developed to monitor CAAL in environmental matrices; however, none approach the above targeted sensitivity levels. Recently a liquid chromatographic method employing column switching was described for the direct determination of CAAL in ground water at a limit of detection (LOD) of 1 ng/mL or at an LOD of 0.1 ng/mL with a concentration step (5). The liquid chromatographic method was unable to distinguish between the *cis-* and *trans-* isomers and required a separate GC with mass spectrometric detection (GC/MS) procedure for confirmation.

This paper describes the development of methods for each isomer of CAAL and CAAC in water and soil. The methods are based upon concentration, cleanup, and derivatization, allowing sensitive detection and confirmation of analytes in each sample by GC/MS using electron impact (EI) ionization with selective ion monitoring (SIM).

EXPERIMENTAL

Instrumentation. A Hewlett-Packard Model 5890 GC and a Model 5971A MSD equipped with a Model G1034B system software (Palo Alto, CA) was used for GC/MS analyses. A J&W Scientific DB-17 capillary column, 20 m x 0.18 mm i.d. with a 0.3-μm film thickness (Folsom, CA) was employed for the determination of CAAL. A J&W Scientific DB-5 capillary column, 30 m x 0.25 mm i.d. with a 0.25-μm film thickness was used for the determination of CAAC.

Standards and Reagents. Standards of CAAL and CAAC isomers were obtained from DowElanco (Test Substance Coordinator, Indianapolis, IN). Isobutyl chloroformate and pyridine were purchased from Aldrich Chemical Company (Milwaukee, WI). *N*-Methyl-*N*-(*t*-butyldimethylsilyl)trifluoroacetamide (MTBSTFA) was obtained from Pierce (Rockford, IL). All solvents were of HPLC grade or better and purchased from Fisher Scientific (Pittsburgh, PA). Hydrochloric acid and sodium hydroxide solutions were of ACS reagent grade and purchased from Fisher Scientific. Reagent grade sodium chloride and anhydrous sodium sulfate were obtained from Fisher Scientific.

Materials. Glassware used to concentrate large volume extracts of CAAL consisted of a 50-mL Erlenmeyer flask and a micro Snyder distillation column, each with a 19/22 ground-glass joint (Kontes, Vineland, NJ). Solid-phase extraction (SPE) columns containing 1 g of silica gel were purchased from J.T. Baker, Inc. (Phillipsburg, NJ). Strong anion-exchange (quaternary amine) SPE columns containing 1 g of packing were obtained from J.T. Baker, Inc.

Surface water and soil used in this study were obtained from a field site near Immokalee, Florida. The soil is classified as a Myakka sand (sandy, siliceous, hyperthermic Aeric Haplaquods) by the USDA-NRCS.

Preparation of CAAL and CAAC Fortification Standards. All standards for fortification were prepared in acetone. A 1000-µg/mL stock solution was prepared for each CAAL isomer. A second stock solution was prepared containing 10 µg/mL of each CAAL isomer. Fortification standards were prepared from the 10-µg/mL solution to give concentrations of each CAAL isomer ranging from 4.0 to 200 ng/mL. Based upon the use of a 40-mL water and 10-g soil sample size, control samples fortified with 1 mL of the appropriate solution resulted in concentrations of each CAAL isomer ranging from 0.10 to 5.0 ng/mL for water and 0.40 to 20 ng/g for soil. Fortification of soil samples at concentrations above 20 ng/g were carried out using the appropriate volume of the 10-µg/mL solution.

A 1000-µg/mL stock solution was prepared for each CAAC isomer. A second stock solution was prepared containing 10 µg/mL of each CAAC isomer. Fortification standards were prepared from the 10-µg/mL solution to give concentrations of each CAAC isomer ranging from 2.0 to 200 ng/mL. Based upon the use of a 40-mL water and 10-g soil sample size, control samples fortified with 1 mL of the appropriate solution resulted in concentrations of each CAAC isomer ranging from 0.05 to 5.0 ng/mL for water and 0.20 to 20 ng/g for soil. Fortification of soil samples at concentrations above 20 ng/g were carried out using the appropriate volume of the 10-µg/mL solution.

Preparation of CAAL and CAAC Calibration Standards. All standards for calibration of CAAL were prepared in hexane. A stock solution containing 10 µg/mL of each CAAL isomer was prepared from the above 1000-µg/mL solutions. Calibration standards were prepared from the 10-µg/mL solution to give concentrations of each CAAL isomer ranging from 2.0 to 400 ng/mL. Based upon a 40-mL water and 10-g soil sample concentrated to a final volume of 1 mL, calibration standards represented an equivalent sample concentration of each CAAL isomer ranging from 0.05 to 10 ng/mL for water and 0.2 to 40 ng/g for soil. A 1-mL aliquot of each calibration standard was derivatized along with samples as described later.

All standards for calibration of CAAC were prepared in a solution containing 0.1% acetic acid and 5% acetone in isooctane (volume %). A stock solution containing 10 µg/mL of each CAAC isomer was prepared from the above 1000-µg/mL solutions. Calibration standards were prepared from the 10-µg/mL solution to give concentrations of each CAAC isomer ranging from 2.0 to 500 ng/mL. Based upon a 40-mL water and 10-g soil sample concentrated to a final volume of 0.5 mL, calibration standards represented an equivalent sample concentration of each CAAC isomer ranging from 0.025 to 6.25 ng/mL for water and 0.1 to 25 ng/g for soil. A 0.5-mL aliquot of each calibration standard was derivatized along with samples by the addition of 25 µL of MTBSTFA.

Methods for the Determination of CAAL in Water and Soil

A flowchart for the determination of CAAL in water and soil is shown in Figure 2. With the exception of the requirement for acid extraction of the soil and passing the extract over a strong anion-exchange SPE column, both water and soil determinations followed the same procedure. After the addition of 10 μL of 1-propanol and 15 g of sodium chloride, residues of CAAL were partitioned from the water or soil extract with methyl-*t*-butyl ether (MTBE). The addition of 1-propanol significantly decreased evaporative losses of CAAL in subsequent concentration steps. CAAL residues in MTBE were passed through a silica gel SPE column and transferred to a 50-mL Erlenmeyer flask. After the addition of 3 mL of hexane and 0.1 g of sodium sulfate, a micro Snyder column was attached to the flask. The flask was placed on a hot plate and the contents were concentrated to a volume of approximately 1 mL. The flask contents were transferred to an 8-mL vial, placed in a water bath at ambient temperature and concentrated to 0.5 mL under a gentle flow of nitrogen that was adjusted to minimize solvent disturbance. The final volume was adjusted to 1 mL with hexane.

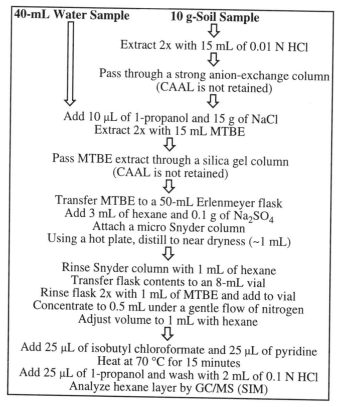

Figure 2. Flowchart of methods for the determination of CAAL in water and soil.

Derivatization of CAAL. Derivatization of CAAL samples and calibration standards were performed in 8-mL vials. After the addition of 25 μL of isobutyl chloroformate and 25 μL of pyridine, the vial was capped and heated at 70 °C for 15 minutes. The vial was allowed to cool to ambient temperature and 25 μL of 1-propanol was added to react with excess reagent. The vial contents were washed with 2 mL of 0.1 N hydrochloric acid and the hexane layer was transferred to an injection vial.

GC/MS Conditions for the Determination of CAAL in Water and Soil. All samples and calibration standards were analyzed for CAAL by GC/MS using a DB-17 capillary column as described previously. Helium was used as the carrier gas at a head pressure of 100 kPa, which gave a linear velocity of approximately 40 cm/sec at an oven temperature of 130 °C. A 2-μL splitless injection was performed at an injector temperature of 230 °C. A splitter delay of 0.7 minutes was used and the splitter and septum purge were set at 50 and 1 mL/min, respectively. The column was held at an initial temperature of 65 °C for 1 minute and ramped at 5 °C/min to 150 °C, then at 20 °C/min to a final temperature of 260 °C. The GC/MS transfer line was maintained at 280 °C. The GC/MS was operated under EI ionization and tuned using the maximum sensitivity autotune program provided with the system software. The electron multiplier was set at 200 volts above the tune voltage. The GC/MS was operated in the SIM mode, monitoring ions at m/z 136 and 75 with a dwell time of 100 msec.

Methods for the Determination of CAAC in Water and Soil. A flowchart of methods for the determination of CAAC in water and soil is shown in Figure 3. With the exception of the requirement for acidified acetone extraction of the soil, concentration to remove acetone, addition of water and adjustment to pH 7, both water and soil determinations followed the same procedure. Residues of CAAC were partitioned onto a strong anion-exchange SPE column and eluted with 5 mL of 0.1 N hydrochloric acid. After the addition of 100 μL of 2.0 N hydrochloric acid and 2 g of sodium chloride, residues of CAAC were partitioned into MTBE. The residues of

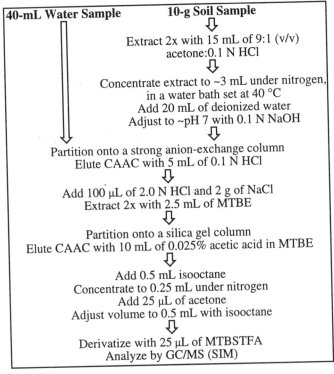

Figure 3. Flowchart of methods for the determination of CAAC in water and soil.

CAAC were then partitioned from MTBE onto a silica gel SPE column and eluted with 10 mL of a solution containing 0.025% acetic acid by volume in MTBE. After the addition of 0.5 mL of isooctane, the eluent was concentrated to 0.25 mL under a flow of nitrogen, in a water bath at ambient temperature. After the addition of 25 μL of acetone, the final volume was adjusted to 0.5 mL with isooctane. The sample was derivatized with 25 μL of MTBSTFA and transferred to an injection vial.

GC/MS Conditions for the Determination of CAAC in Water and Soil. All samples and calibration standards were analyzed for CAAC by GC/MS using a DB-5 capillary column as described previously. Helium was used as the carrier gas at a head pressure of 50 kPa, which gave a linear velocity of approximately 40 cm/sec at an oven temperature of 140 °C. A 1-μL splitless injection was performed at an injector temperature of 230 °C. A splitter delay of 0.5 minutes was used and the splitter and septum purge were set at 50 and 1 mL/min, respectively. The column was held at an initial temperature of 45 °C for 1 minute and ramped at 20 °C/min to 220 °C. The GC/MS transfer line was maintained at 300 °C. The GC/MS was operated under EI ionization and tuned using the maximum sensitivity autotune program provided with the system software. The electron multiplier was set at 200 volts above the tune voltage. The GC/MS was operated in the SIM mode, monitoring ions at m/z 163 and 165 with a dwell time of 100 msec.

RESULTS AND DISCUSSION

Evaluation of Derivatives for use in the Determination of CAAL

A number of derivatization approaches were evaluated for use in the determination of CAAL by GC/MS employing EI ionization. Attempts to utilize trimethylsilyl (TMS) and *t*-butyldimethylsilyl (TBDMS) derivatives were unsuccessful. The dimethyl-pentafluorophenylsilyl derivative (*6*) was readily formed but lacked sensitivity owing to the major fragmentation route through loss of chlorine. The latter derivative gave a strong M-35 ion abundance at m/z 281, susceptible to high background from column bleed and vacuum grease. The condensation of CAAL with triethyl orthoacetate produced derivatives with the necessary sensitivity and chromatographic behavior but impurities in the reagent prevented application to trace level analysis. No evidence of a Claisen rearrangement as described in the literature for allylic alcohols was observed (*7*). Acylations including the pentafluorobenzoyl derivative of CAAL lacked sensitivity and selectivity. The latter derivative has been described (*5*) for the confirmation of CAAL by GC/MS employing negative chemical ionization (NCI). Alkyl chloroformates have been demonstrated to be useful for the derivatization of various aminoalcohols and hydroxycarboxylic acids under mild conditions (*8,9*). The methyl, butyl, isobutyl, and hexyl chloroformates were examined for derivatization of CAAL. Based upon the chromatographic performance, and sensitive and selective detection of the derivatives, isobutyl chloroformate was selected to form the corresponding *cis*- and *trans*-3-chloroallyl isobutyl carbonates (Figure 4). The mass spectra of the *cis*- and *trans*- derivatives were identical. The mass spectrum of the

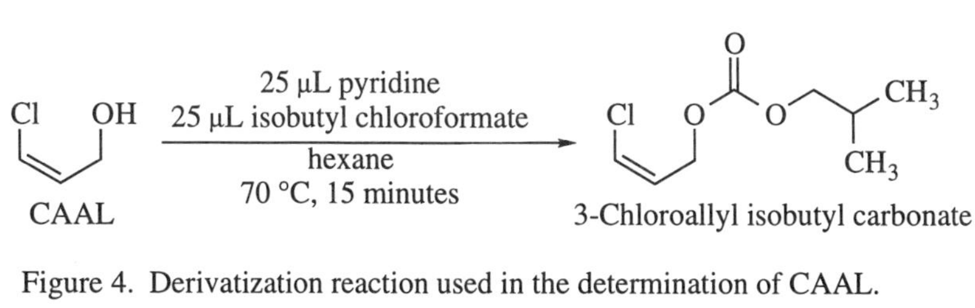

Figure 4. Derivatization reaction used in the determination of CAAL.

cis- derivative is shown in Figure 5. The m/z 136 ion resulting from loss of 2-methyl-2-propene was chosen for quantitation and the m/z 75 ion was selected for confirmation. Both ions retain the chlorine functionality of the CAAL.

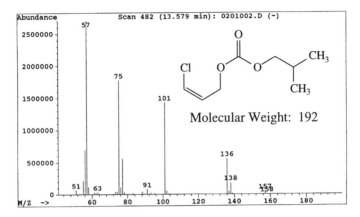

Figure 5. Mass spectrum of the isobutyl chloroformate derivative of *cis*-3-chloroallyl alcohol (*cis*-3-chloroallyl isobutyl carbonate).

Evaluation of Derivatives for use in the Determination of CAAC. Previous methodology for CAAC employed the use of bis(trimethylsilyl)acetamide (BSA) to form the trimethylsilyl esters (TMSE) of CAAC. The use of BSA was limited to 4 µL to minimize interference from excess reagent. To improve sensitivity and separation of derivatives from excess reagent, the use of MTBSTFA to form the *t*-butyldimethylsilyl esters (TBDMSE) of CAAC was examined (Figure 6). Mass spectra of equivalent amounts of the TMSE and TBDMSE derivatives of *trans*-CAAC are shown in Figures 7 and 8, respectively. Abundances of the M-57 ion from the EI mass spectra of TBDMSE derivatives of CAAC were found to be 3 times greater than the corresponding M-15 ion of the TMSE derivatives. No interference from excess reagent was observed in the derivatization of CAAC using MTBSTFA. Reports in the literature suggest TBDMS derivatives in general exhibit greater stability than the TMS derivatives (*10*). Mass spectra of TBDMSE derivatives of *cis*- and *trans*-CAAC differ noticeably in their lower mass fragment ions. The most abundant fragment of both, however, as shown in Figure 8, results from loss of the *t*-butyl radical. This ion at m/z 163 was chosen for quantitation, and it's chlorine-37 isotope at m/z 165 was selected for confirmation.

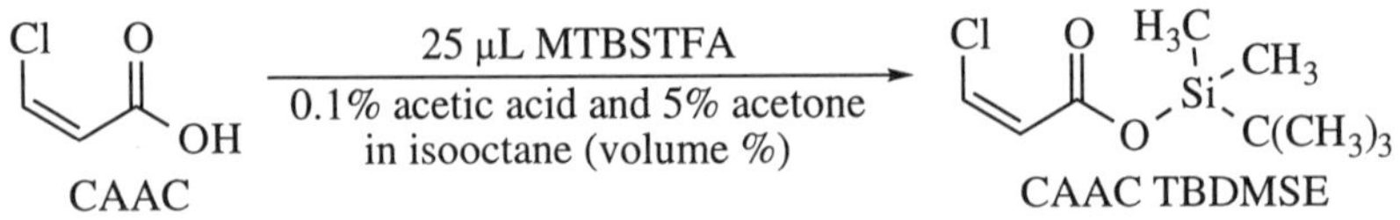

Figure 6. Derivatization reaction used in the determination of CAAC.

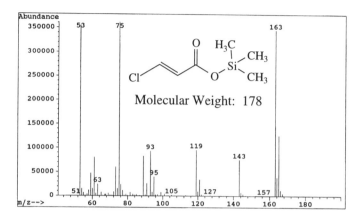

Figure 7. Mass spectrum of the BSA derivative of *trans*-3-chloroacrylic acid (*trans*-3-chloroacrylic acid trimethylsilyl ester).

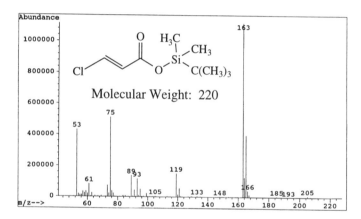

Figure 8. Mass spectrum of the MTBSTFA derivative of *trans*-3-chloroacrylic acid (*trans*-3-chloroacrylic acid *t*-butyldimethylsilyl ester).

Validation of Methods for the Determination of CAAL

Methods for the determination of residues of CAAL in water and soil were validated by fortification of control samples at concentrations ranging from 0.10 to 5.2 ng/mL in water and 0.42 to 2100 ng/g in soil. A six-point calibration containing standard concentrations ranging from 2.1 to 420 ng/mL was performed in duplicate with each validation set. The typical correlation coefficient for the regression equation describing the m/z 136 ion response as a function of the concentration was greater than 0.999 for both CAAL isomers. Concentrations of CAAL in samples were calculated using the m/z 136 ion response and the regression equation derived from calibration standards. A typical chromatogram of a derivatized standard containing 4.2 ng/mL of each CAAL isomer, equivalent to 0.10 ng/mL in water and 0.42 ng/g in soil is shown in Figure 9. The chromatogram displays both the m/z 136 quantitation ion and the m/z 75 confirmation ion responses. Confirmation of CAAL residues in both water and soil samples was based upon retention time and a confirmation ratio, defined as the ratio of the quantitation ion response over the confirmation ion response. The criteria for confirmation of CAAL in water and soil samples required a ±15% agreement of the sample confirmation ratio with the average of the confirmation ratios for the respective calibration standards.

Typical chromatograms of a control soil and a control soil fortified at 0.42 ng/g with each CAAL isomer are shown in Figures 10 and 11. Arrows on the chromatogram of the control soil indicate the elution times of CAAL, no detectable peaks were observed at the expected retention times of each CAAL isomer for both ions monitored. Recoveries of CAAL in the fortified soil were calculated at 88 and 89% for the *cis*- and *trans*- isomers, respectively. Recoveries for the determination of *cis*- and *trans*-CAAL in water and soil are summarized in Tables I and II. Recoveries of 19 fortified surface water samples averaged 83% for the *cis*- and 84% for the *trans*-isomers, each with a standard deviation (SD) of 6%. Recoveries of 23 fortified soil samples averaged 86% with a SD of 8% for the *cis*- and 89% with a SD of 7% for the *trans*- isomers. Soil samples fortified at levels above 21 ng/g were diluted 100-fold with hexane after derivatization to fall within the concentration range of the calibration standards. Confirmation ratios for CAAL in all recovery samples were within ±15% of the average confirmation ratio for the respective standards.

Following published guidelines (*11*), the LOQ and LOD were calculated using the standard deviation of the concentrations found in samples fortified at 0.1 ng/mL for surface waters and at 0.42 ng/g for soils. The LOQ and LOD were calculated as 10 and 3 times the standard deviation, respectively. The calculated LOQ for each CAAL isomer in surface water and soil were 0.08 ng/mL and 0.4 ng/g. The calculated LOD for each CAAL isomer is surface water and soil were 0.02 ng/mL and 0.1 ng/g. In support of the calculated LOD, duplicate surface water and soil controls were fortified with each CAAL isomer at levels of 0.02 ng/mL and 0.08 ng/g. Both CAAL isomers were detected in all samples fortified at the LOD.

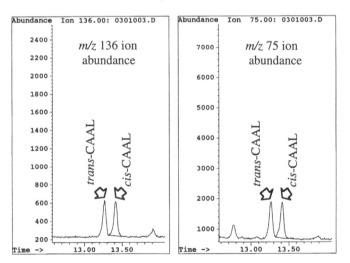

Figure 9. Typical chromatogram of a standard containing 4.2 ng/mL of each CAAL isomer, equivalent to 0.1 ng/mL in water and 0.42 ng/g in soil.

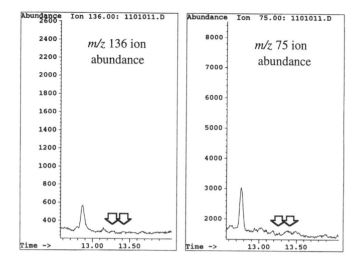

Figure 10. Typical chromatogram of a control soil sample for the determination of CAAL isomers.

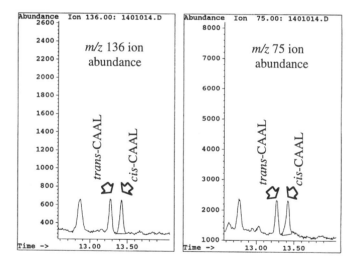

Figure 11. Typical chromatogram of a 0.42 ng/g fortified control soil sample for the determination of CAAL.

Storage Stability of CAAL in Water and Soil. A storage stability study was conducted for samples of surface water and soil fortified with each CAAL isomer at levels of 1 ng/mL and 1 ng/g. Water samples were stored refrigerated at 4 °C and soil samples were stored frozen at -20 °C. Periodically, samples were analyzed in triplicate for each CAAL isomer. Fortified surface waters showed no losses of CAAL after 7 days of refrigerated storage. After 14 and 30 days, losses of 10 and 50%, respectively were observed. No losses of CAAL were observed in fortified soils after 61 days of storage.

Table I. Summary of recoveries for the determination of *cis*- and *trans*-CAAL in water

| Fortification Level | | Percent Recovery | | | |
| | | *cis*-CAAL | | *trans*-CAAL | |
ng/mL	Trials	Average	SD[a]	Average	SD[a]
0.10	8	83	7	83	7
0.26	2	84	4	84	4
0.52	3	87	3	87	3
1.0	2	81	3	84	3
2.6	2	90	3	94	4
5.2	2	76	1	80	1
Total	19	83[b]	6	84[b]	6

[a]Standard deviation of the average.
[b]Average of all recovery trials.

Table II. Summary of recoveries for the determination of *cis*- and *trans*-CAAL in soil

| Fortification Level | | Percent Recovery | | | |
| | | *cis*-CAAL | | *trans*-CAAL | |
ng/g	Trials	Average	SD[a]	Average	SD[a]
0.42	9	88	10	90	9
1.0	2	94	5	94	1
2.1	2	79	6	82	6
10	2	82	7	84	7
21	2	92	3	95	4
104	2	78	1	83	1
520	2	81	4	84	5
2080	2	88	6	90	6
Total	23	86[b]	8	89[b]	7

[a]Standard deviation of the average.
[b]Average of all recovery trials.

Precautionary Note. Prior to validation of the CAAL methods, a *m/z* 75 interference was observed at the retention time of the *cis*-CAAL derivative. The interference was identified as naphthalene by spectral and retention matches to a standard. Although it is unlikely that such levels of naphthalene would be found in environmental samples, some commercial cleansers should not be used in a trace level laboratory. No interference was observed when use of the cleanser was discontinued.

Validation of Methods for the Determination of CAAC

Methods for the determination of residues of CAAC in water and soil were validated by fortification of control samples at concentrations ranging from 0.05 to 5.0 ng/mL in water and 0.20 to 2000 ng/g in soil. A six-point calibration containing standard concentrations ranging from 2.0 to 500 ng/mL was performed in duplicate with each validation set. The typical correlation coefficient for the regression equation describing the *m/z* 163 ion response as a function of the concentration was greater than 0.999 for both CAAC isomers. Concentrations of CAAC in samples were calculated using the *m/z* 163 ion response and the regression equation derived from calibration standards. A typical chromatogram of a derivatized standard containing 4.0 ng/mL of each CAAC isomer, equivalent to 0.05 ng/mL in water and 0.20 ng/g in soil is shown in Figure 12. The chromatogram displays both the *m/z* 163 quantitation

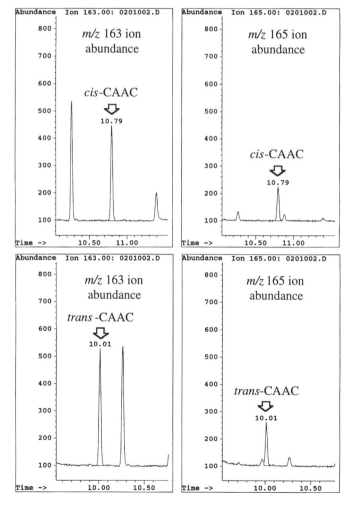

Figure 12. Typical chromatogram of a standard containing 4.0 ng/mL of each CAAC isomer, equivalent to 0.05 ng/mL in water and 0.20 ng/g in soil.

ion and the m/z 165 confirmation ion responses. Confirmation of CAAC residues in both water and soil samples was based upon retention time and a confirmation ratio, defined as the ratio of the confirmation ion response over the quantitation ion response. The criteria for confirmation of CAAC in water and soil samples required a ±15% agreement of the sample confirmation ratio with the average of the confirmation ratios for the respective calibration standards.

Typical chromatograms of a control soil and a control soil fortified at 0.20 ng/g with each CAAC isomer are shown in Figures 13 and 14. A response was observed at the expected retention times of each CAAC isomer representing approximately 1/10 of the respective 0.20 ng/g equivalent standard. Reagent blanks carried out with sample sets gave similar results, indicating a procedural source of contamination. The low level contamination was not considered a detriment to the validation. All

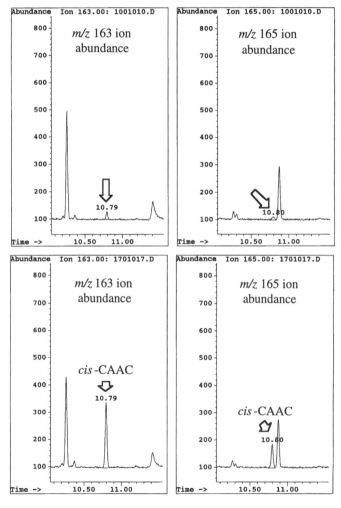

Figure 13. Typical chromatograms of a control soil (top) and a 0.20 ng/g fortified soil (bottom) for the determination of *cis*-CAAC.

recovery calculations were corrected for the average CAAC response of control samples. Recoveries of CAAC in the fortified example were calculated at 84 and 82% for the *cis-* and *trans-* isomers, respectively. Recoveries for the determination of *cis-* and *trans*-CAAC in water and soil are summarized in Tables III and IV. Recoveries of 18 fortified surface water samples averaged 91% for the *cis-* and 92% for the *trans-* isomers, each with a SD of 7%. Recoveries of 25 fortified soil samples averaged 80% with a SD of 7% for the *cis-* and 84% with a SD of 6% for the *trans-*isomers. Soil samples fortified at levels above 20 ng/g were diluted 100-fold with isooctane after derivatization to fall within the concentration range of the calibration standards. Confirmation ratios for CAAC in all recovery samples were within ±15% of the average confirmation ratio for the respective standards.

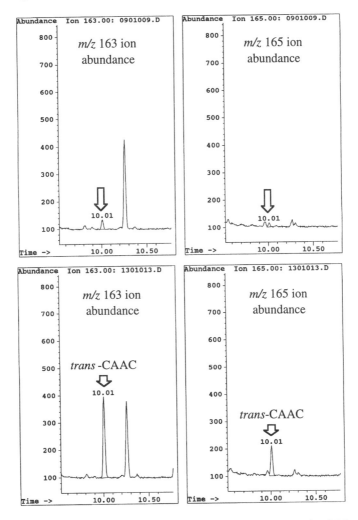

Figure 14.   Typical chromatograms of a control soil (top) and a 0.20 ng/g fortified soil (bottom) for the determination of *trans*-CAAC.

The LOQ and LOD were calculated using the standard deviation of the concentrations found in samples fortified at 0.05 ng/mL for surface waters and at 0.20 ng/g for soils. The LOQ and LOD were calculated as 10 and 3 times the standard deviation. The calculated LOQ for each CAAC isomer in surface water and soil were 0.03 ng/mL and 0.2 ng/g. The calculated LOD for each CAAC isomer in surface water and soil were 0.009 ng/mL and 0.06 ng/g. In support of the calculated LOD, duplicate surface water and soil controls were fortified with each CAAC isomer at levels of 0.009 ng/mL and 0.06 ng/g. Both CAAC isomers were detected in all samples fortified at the LOD.

Table III. Summary of recoveries for the determination of *cis-* and *trans*-CAAC in water

Fortification Level		Percent Recovery			
		cis-CAAC		*trans*-CAAC	
ng/mL	Trials	Average	SD[a]	Average	SD[a]
0.05	8	89	6	90	6
0.25	2	91	3	92	2
0.50	2	84	13	86	14
1.25	2	97	8	100	8
2.5	2	92	5	94	5
5.0	2	93	13	94	13
Total	18	91[b]	7	92[b]	7

[a]Standard deviation of the average.
[b]Average of all recovery trials.

Storage Stability of CAAC in Water and Soil. A storage stability study was conducted for samples of surface water and soil fortified with each CAAC isomer at levels of 1 ng/mL in water and 1 ng/g in soil. Water samples were stored refrigerated at 4 °C and soil samples were stored frozen at -20 °C. Periodically, samples were analyzed in triplicate for each CAAC isomer. Fortified surface waters showed no losses of CAAC after 60 days of refrigerated storage. No losses of CAAC were observed in fortified soils after 61 days of storage.

Precautionary Notes. The procedural contamination encountered during this study at approximately 1/10 of the targeted LOQ was not considered a significant detriment to the method validation. Care should be taken to prevent cross-contamination of reagents used in the procedure, particularly when high levels of CAAC are involved. Equipment used in the determination of CAAC should be rinsed with acidified acetone solution prior to reuse. Strong anion-exchange and silica gel SPE columns should be evaluated prior to use for optimum method performance.

CONCLUSIONS

The methods presented in this paper are sensitive and selective for the determination of residues of CAAL and CAAC in water and soil. They are currently in use for the determination of CAAL and CAAC in ground and surface water, and soil, toward assessment of the environmental fate of 1,3-D. Water, transported and stored refrigerated, should be analyzed for CAAL within 14 days of sampling to prevent losses.

Table IV. Summary of recoveries for the determination of *cis*- and *trans*-CAAC in soil

| Fortification Level | | Percent Recovery | | | |
| | | *cis*-CAAC | | *trans*-CAAC | |
ng/g	Trials	Average	SD[a]	Average	SD[a]
0.20	9	83	11	87	9
1.0	2	74	1	81	0
2.0	2	77	6	80	7
5.0	2	84	3	85	3
10	2	78	1	82	1
20	2	77	0	80	1
100	2	82	2	84	2
500	2	76	5	80	2
2000	2	77	4	80	4
Total	25	80[b]	7	84[b]	6

[a]Standard deviation of the average.
[b]Average of all recovery trials.

References

1. Castro, C. E.; Belser, N. O. *J. Agr. Food Chem.* **1966**, *14*, 69-70.
2. Belser, N. O.; Castro, C. E. *J. Agr. Food Chem.* **1971**, *19*, 23-26.
3. van Dijk, H. *Agro-Ecosystems* **1974**, *1*, 193-204.
4. Roberts, T. R.; Stoydin, G. *Pestic. Sci.* **1976**, *7*, 325-335.
5. Hogendoorn, E. A.; de Jong, A. P. J. M.; van Zoonen, P.; Brinkman, U. A. Th. *J. Chromatogr.* **1990**, *511*, 243-256.
6. Francis, A. J.; Morgan, E. D.; Poole, C. F. *J. Chromatogr.* **1978**, *161*, 111-117.
7. Johnson, W. S.; Werthemann, L.; Bartlett, W. R.; Brocksom, T. J.; Li, T.; Faulkner, D. J.; Petersen, M. R. *J. Am. Chem. Soc.* **1970**, *92*, 741-743.
8. Husek, P.; Huang, Z-H.; Sweeley, C. C. *Anal. Chim. Acta* **1992**, *259*, 185-192.
9. Butz, S.; Stan, H. -J. *J. Chromatogr.* **1993**, *643*, 227-238
10. Mawhinney, T. P.; Madson, M. A. *J. Org. Chem..* **1982**, *47*, 3336-3339.
11. Keith, L. H.; Crummett, W.; Deegan, J., Jr.; Libby, R. A.; Taylor, J. K.; Wentler, G. *Anal. Chem.* **1983**, *55*, 2210-2218.

INDEXES

Author Index

Affiliation Index

Subject Index

Highlights from ACS Books

Chemical Research Faculties, An International Directory
1,300 pp; clothbound ISBN 0–8412–3301–2

College Chemistry Faculties 1996, Tenth Edition
300 pp; paperback ISBN 0–8412–3300–4

Visualizing Chemistry: Investigations for Teachers
By Julie B. Ealy and James L. Ealy
456 pp; paperback ISBN 0–8412–2919–8

Principles of Environmental Sampling, Second Edition
Edited by Lawrence H. Keith
700 pp; clothbound ISBN 0–8412–3152–4

Enough for One Lifetime: Wallace Carothers, Inventor of Nylon
By Matthew E. Hermes
364 pp; clothbound ISBN 0–8412–3331–4

Peptide-Based Drug Design
Edited by Michael D. Taylor and Gordon Amidon
650 pp; clothbound ISBN 0–8412–3058–7

Attenuated Total Reflectance Spectroscopy of Polymers: Theory and Practice
By Marek W. Urban
232 pp; clothbound ISBN 0–8412–3348–9

Teaching General Chemistry: A Materials Science Companion
By Arthur B. Ellis, Margaret J. Geselbracht, Brian J. Johnson, George C. Lisensky,
and William R. Robinson
576 pp; paperback ISBN 0–8412–2725–X

Understanding Medications: What the Label Doesn't Tell You
By Alfred Burger
220 pp; clothbound ISBN 0–8412–3210–5; paperback ISBN 0–8412–3246–6

For further information contact:
American Chemical Society
Customer Service and Sales
1155 Sixteenth Street, NW
Washington, DC 20036
Telephone 800–227–9919
202–776–8100 (outside U.S.)
The ACS Publications Catalog is available on the Internet at
http://pubs.acs.org/books

Bestsellers from ACS Books

The ACS Style Guide: A Manual for Authors and Editors
Edited by Janet S. Dodd
264 pp; clothbound ISBN 0–8412–0917–0; paperback ISBN 0–8412–0943–X

Writing the Laboratory Notebook
By Howard M. Kanare
145 pp; clothbound ISBN 0–8412–0906–5; paperback ISBN 0–8412–0933–2

Career Transitions for Chemists
By Dorothy P. Rodmann, Donald D. Bly, Frederick H. Owens, and Anne-Claire Anderson
240 pp; clothbound ISBN 0–8412–3052–8; paperback ISBN 0–8412–3038–2

Chemical Activities (student and teacher editions)
By Christie L. Borgford and Lee R. Summerlin
330 pp; spiralbound ISBN 0–8412–1417–4; teacher edition, ISBN 0–8412–1416–6

Chemical Demonstrations: A Sourcebook for Teachers, Volumes 1 and 2, Second Edition
Volume 1 by Lee R. Summerlin and James L. Ealy, Jr.
198 pp; spiralbound ISBN 0–8412–1481–6
Volume 2 by Lee R. Summerlin, Christie L. Borgford, and Julie B. Ealy
234 pp; spiralbound ISBN 0–8412–1535–9

From Caveman to Chemist
By Hugh W. Salzberg
300 pp; clothbound ISBN 0–8412–1786–6; paperback ISBN 0–8412–1787–4

The Internet: A Guide for Chemists
Edited by Steven M. Bachrach
360 pp; clothbound ISBN 0–8412–3223–7; paperback ISBN 0–8412–3224–5

Laboratory Waste Management: A Guidebook
ACS Task Force on Laboratory Waste Management
250 pp; clothbound ISBN 0–8412–2735–7; paperback ISBN 0–8412–2849–3

Reagent Chemicals, Eighth Edition
700 pp; clothbound ISBN 0–8412–2502–8

Good Laboratory Practice Standards: Applications for Field and Laboratory Studies
Edited by Willa Y. Garner, Maureen S. Barge, and James P. Ussary
571 pp; clothbound ISBN 0–8412–2192–8

For further information contact:
American Chemical Society
1155 Sixteenth Street, NW ◆ Washington, DC 20036
Telephone 800–227–9919 ◆ 202–776–8100 (outside U.S.)
The ACS Publications Catalog is available on the Internet at
http://pubs.acs.org/books